Creating Neighbourhoods and Places in the Built Enviro

THE BUILT ENVIRONMENT SERIES OF TEXTBOOKS (BEST)

Executive Editor:	Professor Tony Collier, Dean, Faculty of the Built Environment, University of Central England, Birmingham, UK
Co-ordinating Editor:	David Burns, Faculty of the Built Environment, University of Central England, Birmingham, UK
Assistant Editor:	Jean Bacon, Faculty of the Built Environment, University of Central England, Birmingham, UK

ADVISORY BOARD:

James Armstrong *Visiting Professor, Faculty of Technology, Kingston University*

David Cormican *Deputy Director, North West Institute of Further and Higher Education, Londonderry, Northern Ireland*

Bryan Jefferson *Architectural Advisor to the Secretary of State, Department of National Heritage*
Visiting Professor, Sheffield University, Faculty of Architectural Studies

Howard Land *Professional Training Consultant, RICS*

Alan Osborne *Chairman, Construction Industry Standing Conference (CISC)*

John Tarn *Professor of Architecture, Pro-Vice Chancellor, University of Liverpool*

Alan Wenban-Smith *Principal consultant, Segal Quince Wickstead Ltd (Engineering and Management Consultants)*

This series of textbooks responds to changes that are occurring throughout the construction industry and in higher and further education. It focuses on aspects of the curriculum that are common to all professions in the built environment. The principal aim of BEST is to provide texts that are relevant to more than one course and the texts therefore address areas of commonality in an original and innovative way. Learning aids in the texts such as chapter objectives, checklists, and workpieces will appeal to all students.

OTHER TITLES IN THE SERIES:

Design, Technology and the Development Process in the Built Environment
Management and Business Skills in the Built Environment
Collaborative Practice in the Built Environment
Legal Frameworks for the Built Environment

CREATING NEIGHBOURHOODS AND PLACES IN THE BUILT ENVIRONMENT

EDITED BY DAVID CHAPMAN

Faculty of the Built Environment,

University of Central England, UK

E & FN SPON

An Imprint of Chapman & Hall

London · Weinheim · New York · Tokyo · Melbourne · Madras

**Published by E & FN Spon, an imprint of Chapman & Hall,
2–6 Boundary Row, London SE1 8HN, UK**

Chapman & Hall, 2–6 Boundary Row, London SE1 8HN, UK

Chapman & Hall, GmbH, Pappelallee 3, 69469 Weinheim, Germany

Chapman & Hall USA, 115 Fifth Avenue, New York, NY 10003, USA

Chapman & Hall Japan, ITP-Japan, Kyowa Building, 3F, 2-2-1 Hirakawacho,
Chiyoda-ku, Tokyo 102, Japan

Chapman & Hall Australia, 102 Dodds Street, South Melbourne, Victoria 3205,
Australia

Chapman & Hall India, R. Seshadri, 32 Second Main Road, CIT East, Madras
600 035, India

First edition 1996

© 1996 E & FN Spon

Typeset in 11/14 Caslon by Saxon Graphics Ltd, Derby
Printed in Great Britain by the Alden Press, Osney Mead, Oxford

ISBN 0 419 20930 1

A catalogue record for this book is available from the British Library

Library of Congress Catalog Card Number: 96–67189

∞ Printed on permanent acid-free text paper, manufactured in accordance
with ANSI/NISO Z39.48-1992 and ANSI/NISO Z39.48-1984 (Permanence of
Paper).

CONTENTS

LIST OF CONTRIBUTORS

Martin Bradshaw
Former Director
Civic Trust
London

David Chapman
Head of School of Planning
Faculty of the Built Environment
University of Central England
Birmingham

John Donovan
Freelance Urban Designer

Dr Peter J. Larkham
Senior Lecturer
School of Planning
Faculty of the Built Environment
University of Central England
Birmingham

Tom Muir
Head of Foundation Studies
Faculty of the Built Environment
University of Central England
Birmingham

Kevin Murray
Director
EDAW CR Planning
Glasgow

Dick Pratt
Senior Lecturer
School of Planning
Faculty of the Built Environment
University of Central England
Birmingham

Les Sparks
Director of Planning and
    Architecture
Birmingham City Council

Colin Wood
Senior Lecturer
School of Planning
Faculty of the Built Environment
University of Central England
Birmingham

L. John Wright
Research Associate
School of Geography
University of Birmingham

ACKNOWLEDGEMENTS

Many people have assisted in the development of this book. I would especially like to thank Professor Tony Collier for inviting me to lead it and the chapter authors for their dedication in sometimes very difficult circumstances.

The support of Beryl Stanton in preparing the manuscript, Steve Roddie and Sally Jones in producing illustrations, and Andrew Ford's help with indexing are greatly appreciated. Special thanks must go to Peter Larkham for his advice on editing and collaboration on a range of research projects which have contributed to the development of this book. Joe Holyoak's work has contributed to and inspired several sections of the book, and Maurice Ingram also contributed the Eco House case study.

David Chapman

INTRODUCTION

This book follows directly from *Design Technology and the Development Process*, which explored the fundamental generators and contextual issues that influence the nature of the built environment. It extends the four main themes of that volume beyond small groups of buildings to the nature of settlements and how buildings, spaces and human activities combine to create lively and enjoyable neighbourhoods and places.

The four main themes of this book are:

- social, political and economic forces, and the way people interact in the creation of buildings and places;
- how society and technology change and the way these changes affect the environment;
- how we intervene with the natural world to create settlements, to modify microclimates and create spaces which are comfortable for people to use;
- the nature of problem solving activities in the built environment, particularly creativity in the management of settlements and places.

The book addresses important issues about which all young built environment professionals should know, regardless of their future career specialisms. It is aimed primarily at undergraduate students on built environment courses but it is relevant to a wide range of other disciplines and postgraduates, as well as a range of vocational and environmental programmes.

The development of built environments results from the interaction of a complex and diverse range of forces and the actions of a multitude of individuals and agencies. All students and practitioners interested and involved in these issues need to understand the forces at work and gain insight into the implications of their actions and ways that they may support positive outcomes. This book explores some of the competing interests which exist, their interaction with physical and environmental forces, and the uncertainty about the outcomes of both individual and corporate intervention.

Each chapter examines the forces at work, the form of the resultant buildings, built environments and landscapes, and the nature of their use and vitality. In doing so it seeks to recognize the importance of careful analysis and of innovation and creativity.

ABOUT THIS BOOK

Throughout the book there is a clear focus upon the ways in which the built environment, including the processes of incremental change as well as major building projects, affects the quality of life of local and international communities. It seeks to give an integrating perspective of town and country planning, urban design, architecture and landscape design and to recognize the importance of the wide range of disciplines (for example, engineers, surveyors and estate managers) which contribute to the formation and character of the built environment. Key factors include natural resources, transport and communications, design and conservation, and the ways in which private ownership and public interests interrelate.

OBJECTIVES

The book has some key objectives:

- To explore the range of forces which create and change built environments.
- To appreciate the variety and complexity of these forces.
- To recognize the uncertainty of the outcomes of the interaction between these forces.
- To consider the diversity of possible outcomes over time and understand that interventions may have a range of effects from short to long term.
- To provide a framework for the analysis of places and their cultural, technological and physical achievements.
- To consider how we as individuals and organizations can contribute towards the achievement of positive and creative outcomes in the quality of our living environments.
- To trace, through a variety of places from different periods and cultures, the way problems have been identified and what solutions were employed to respond to them.

STRUCTURE

This book has been organized into three sections.

Part One explores the **development of settlements**, how they adapt to different environments, how they grow, and the character of urban places today.

Part Two examines the **qualities of places**, how people use places, the qualities which give a place variety and vitality, and the environmental and aesthetic qualities which the many enjoyable places possess.

Part Three looks at ways of **building people-friendly places**, including the development of public policies and plans, the representation of private interests, and the ways in which local communities and single individuals can become involved in decision making.

Chapter 1 (Nature and settlement) is concerned with the relationships between the natural world and the form and pattern of settlements. It explores the variety of ways in which settlements evolve and adapt in response to local conditions and considers some of the ways in which future development may adopt more efficient and sustainable forms.

    Chapter 2 (Settlements and growth) continues by considering, how settlements develop, grow and sometimes decline. It explores through historical examples some of the characteristics of changing settlements, in both growth and decline, and seeks to identify common patterns of activity and change and to understand some of the underlying reasons for the developing form and use of different places.

    Chapter 3 (Modern urban places) examines the form and character of modern urban places – the diversity of forms and qualities and their causes. Rapid technological change since the industrial revolution combined with the emergence of new approaches to planning and architecture have often resulted in urban landscapes and environments of sometimes great but often poor quality.

What is it that makes some places enjoyable and others not? This section explores the qualities which might be protected or created.

    Chapter 4 (Equity and access) looks at the ways in which built environments affect access to facilities and opportunities as well as the impact of places on the equitability of opportunity for their users. In addition to considering equity today it explores the intergenerational equity of today's actions and considers issues for sustainability of places and buildings.

    Chapter 5 (Variety and vitality) highlights the importance of variety and vitality for the quality of places. It explores the characteristics of places which make them lively and responsive to their users' needs and considers some of the ways in which public policies and private activities have encouraged vitality and the ways in which we can enrich the places we use.

    Chapter 6 (Environment and space) explores some of the aesthetic qualities of places, spaces and diverse environments. It considers the physical form and townscapes of places and how they contribute to the environmental and visual experiences of the people who use them.

It looks at the theoretical background to the appraisal and understanding of places and explores these in practice with reference to some of the criteria which affect the quality of places.

PART THREE CREATING PEOPLE-FRIENDLY PLACES

It is equally important that we appreciate the forces and processes which change and shape places for the future. Only by doing so can we see how our actions and influence contribute to the creation of equitable and enjoyable places.

Chapter 7 (**Public policy and planning**) looks at the ways in which public policies and planning can influence the form and quality of places. The instruments of government may provide national frameworks or local guidance for urban development and management. This chapter explores the motivation underlying formulation of policies and plans and the different levels of participation which might be found, as well as the ways in which democratic involvement can be fostered.

Chapter 8 (**Community action and involvement**) examines how people can be positively involved in shaping their local environment, and the steps some have taken to force their interests and wishes into the public domain. It explores the opportunities and needs for positive community and individual involvement and action. It identifies some of the obstacles to participation, steps which can be promoted by public policy and ways in which communities can exert influence from the grass roots.

Chapter 9 (**Renewal and regeneration**) examines how some places have been developed, renewed and redeveloped, and includes consideration of the utopian aspirations which may have inspired them. It also traces a range of utopian approaches to settlement planning and a variety of responses to the needs for urban renewal and area regeneration. The contrasting implications of dispersal and containment policies and their implications for the characteristics of distinctive places are also raised.

Finally, **Chapter 10** (**Making connections**) seeks to draw together the diverse, complementary and conflicting strands which have been introduced throughout the book. The ability to make connections between diverse forces and factors helps us not only to understand the competing interests at stake in society but also to seek strategies which respond to and respect those interests now and in the future. We should not forget however the profound effect of an inspired creative act!

GETTING THE MOST FROM THIS BOOK

This is both a textbook and a workbook. You can read it and work through the exercises over and over again. Try creating different scenarios and consider the possible changing outcomes. There are no correct answers but the process develops our understanding and insight.

As you develop your understanding and knowledge through this book you will wish to consider drawing on other books in the series. *Design Technology and the Development Process* has been mentioned at the beginning of this introduction. *Legislative Frameworks in the Built Environment* covers much of the detailed legislative context – for example, legal definitions and statutory processes – including Planning Policy Guidance notes in the UK. This provides the back-up understanding and knowledge to implement neighbourhood or area planning. *Collaborative Practice* explains ways in which professionals work together, and the title of another book in the series, *Management and Business Skills*, is self-explanatory.

Together the series as a whole provides a complementary set of books to enable young professionals in the twenty-first century to work together using the specialist skills of their own area of work whilst having a breadth of understanding, knowledge and skill about the environment to operate effectively on a range of tasks, including in multidisciplinary teams.

THE DEVELOPMENT
OF SETTLEMENTS

NATURE AND SETTLEMENT

L. JOHN WRIGHT

Each local region, neighbourhood and place has its own unique and dynamic natural environment. Human settlement evolves, partly in response to the infinite variations in environmental conditions, in order to maximize utility and comfort for inhabitants. From earliest times the human race has endeavoured not only to adapt to natural elements but to modify and control them. In the modern world of accelerating population growth and urbanization, the impacts of settlement on natural systems are ever increasing.

Several fundamental questions arise which are important considerations for any student or professional working in the built environment field:

● How far do natural factors influence the siting, form and buildings of rural settlements?
● How much do natural forces influence the location and development of towns and cities?
● What impacts do settlements have on their natural environment?
● Can such impacts be controlled in order to ensure sustainability of settlements within their natural surroundings?

This chapter discusses these issues and looks for possible answers to these key questions.

After reading this chapter you should be able to:

● understand how far geology, landforms and soils influence rural settlement forms;

● appreciate how siting and development of towns and cities are influenced by geology and landforms;

● recognize ways in which urban development is affected by natural barriers;

● appreciate where settlements have not adapted to environments adequately;

● identify ways in which natural landscapes have been modified or controlled for development purposes;

● understand how weather and climate can affect building forms and settlement;

● assess the overall interactions of settlements with natural environment and be able to consider the issue of their mutual sustainability.

INTRODUCTION

Before proceeding further it is worth clarifying what is meant by 'nature' and 'settlement' in the context of this chapter. **Nature**, or the natural environment, comprises different but interrelated systems. These systems are often studied together as physical geography or within a multi-disciplinary environmental studies or earth science package. They can be simplified into four major systems:

● the **lithosphere**, or land system (geology, landforms and soils);
● the **atmosphere**, or air system (weather, climate and air quality);
● the **hydrosphere**, or water system (seas, surface fresh water, ground water and ice);
● the **biosphere**, or life system (plant and animal organisms).

Settlement can be taken simply to refer to any human habitations or dwellings, but settlements are organized into different patterns or distributions. Rural settlement patterns may be **dispersed** (consisting of single isolated dwellings or farms), **clustered** (with loosely spaced groups of dwellings) or **nucleated** (with closely spaced groups of dwellings). Smaller nucleated settlements with between three and 19 dwellings may be termed **hamlets**. Those with 20 or more dwellings can be referred to as **villages**[1]. **Urban settlements** (towns and cities) are distinguishable from villages not only by virtue of their size, normally consisting of at least several hundred dwellings, but also because of their wider range of functions and services.

FORMS OF RURAL SETTLEMENT

Study any reasonably large-scale map available for a rural area in your own region or country (e.g. 1:50 000, 1:25 000 or 1:10 000). Using the descriptive terms for rural settlement types defined in the introductory section:

● Identify any differences in the overall density of rural settlement in the area covered by the map. Where are the most and least villages, hamlets, clusters and isolated dwellings?

● Distinguish any variations in settlement form in the area concerned. Are there more or larger nucleated settlements in some districts than others? Are some parts of the map more characterized by dispersed settlement than others?

● Draw a sketch map of the area and divide it according to types of rural settlement that you have recognized.

Having defined what we mean by 'nature' and 'settlement' it is possible to begin to consider how they interrelate. Settlements and settlement patterns evolve through complex interactions between the physical (natural), socio-economic and cultural groups of forces. So just how influential are the natural forces regarding the evolution of settlements? To approach answers to this question it is proposed first to examine rural settlement patterns and then analyse aspects of urban forms.

Natural forces operate at different geographical scales. At a regional scale do they help to create contrasts in rural settlement forms, say over some hundreds of kilometres? How far do they exert a local influence regarding general situations of settlements to within a few kilometres? How do they influence the precise siting of settlements perhaps to within a few scores of metres?

Illustrations of rural settlement can be drawn from England and Wales in the UK. Even in such a densely populated, industrialized and urbanized country there are contrasting rural forms of ancient derivation. Historians know that these patterns were largely established by the mediaeval period (AD 1100–1400) and so date from times when the rural economy was based on subsistence agriculture. Do these patterns in any way show varied adaptations by farming communities to natural forces, especially those of relief, geology, soils and climate?

NATURE AND RURAL SETTLEMENT

In England and Wales four broad regional rural settlement zones are recognizable [2] (Figure 1.1):

REGIONAL SCALE

● a middle zone of England dominated by villages with relatively few hamlets or scattered farms;

- a western zone of England characterized by hamlets, small clusters and dispersed homesteads with some villages;
- Wales, parts of the English south-west peninsula and extreme north of England with predominant dispersed settlement with some clusters and hamlets;
- some southern and south-eastern regions of England with a mix of settlement forms including dispersed dwellings, clusters, hamlets and villages.

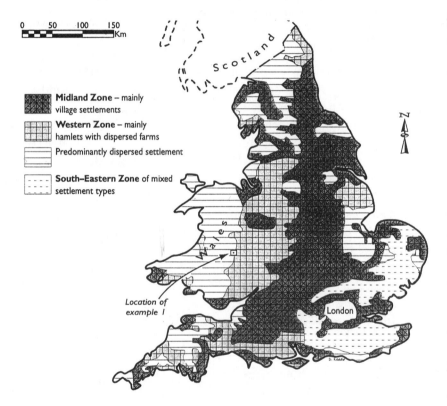

Figure 1.1 Regional rural settlement types in England and Wales.

Analysis of physical geography does indicate that natural forces were very significant formative influences. The middle zone corresponds with some of the most naturally favoured farming regions. It is generally underlain by younger Mesozoic rocks, good soils are usual, lowland (below 200 m) predominates and rainfall is not excessive (generally 550–700 mm per annum). Greater potential for food crops could support a larger population and nucleated villages.

The western zone has more of the older Palaeozoic geology, less consistently good soils, more uplands (above 200 m) and more rainfall (usually above 700 mm even on lowlands). Natural conditions were less conducive to supporting a dense population and any nucleations tended to be smaller hamlets.

In Wales, the south-west and the north, the older Palaeozoic rocks produce chiefly poorer acid soils and hill country with higher rainfall (often 1000 mm on low ground and 2000 mm in hill regions). In Britain this combination of natural factors makes ripening of basic cereal crops difficult and more reliance has to be placed on pastoral farming (sheep and cattle). This farming economy supported a smaller dependent population in scattered rural settlements. In some parts of south and south-east England a mix of settlement forms is perhaps less easily explained in natural terms.

Natural forces are just one set of several very important forces which influence rural settlement. Economic, social and cultural forces are significant, especially concerning how land was farmed. This is evident in England and Wales. The middle zone of nucleated settlement corresponds to rich cropping lands but also to regions where the English (or Anglo-Saxon) settlement and socio-economic organization dominated. Anglo-Saxon settlers came from north Germany in the fifth and sixth centuries. They later developed an agricultural organization which culminated in the manorial system after the Norman Conquest. This system was based on three open arable fields cultivated communally and so centred on the typical nucleated village.

Hamlets and dispersed patterns in the western zone demonstrate the English response to less favourable farming country. Small nucleated settlements were supported by smaller open fields in pairs, rather than threes, and scattered farms often represent mediaeval forest clearings or continued occupation of pre-English settlement sites.

By contrast Wales, the south-west and the northern hill areas retained cultural and economic occupancy by older Celtic-speaking populations. Brythonic Celtic, or Welsh as it has become, survives still in north and west Wales. These Celtic peoples had small scattered plots of arable land with isolated farmhouses, or small clusters of houses, within a tribal area devoted largely to grazing lands. Dispersed patterns resulted and even churches were located in isolation.

Some ancient villages occur in southern and south-eastern regions of England. However, varied physical landscapes and early opportunities for commercial (as opposed to communal subsistence) farming near London were possible factors leading to divergent settlement patterns.

What conclusions emerge from this analysis of regional rural patterns in England and Wales? Here are three to consider:

- Ancient variations in rural settlement pattern persist.
- Different patterns evolved in part as adaptations to variable natural environments and forces.
- Economic and cultural forces were influential, especially those emanating from the differences in organization of subsistence farming.

Brian Roberts, in his book *Rural Settlement in Britain*, summarizes the important role of natural forces by advising his reader: 'It is well to appreciate the extent to which physical factors can operate indirectly on settlement through the intermediary of economic activity'[3].

LOCAL SCALE

Interactions between natural forces and rural settlement evolution can be appreciated even more by closer analysis of a local area. This is especially so where there are obvious contrasts in physical environment linked with what has been historically a cultural transitional zone. Such areas exist along the borders between England and Wales. One has been selected for detailed study in Example 1.1.

EXAMPLE 1.1

LOCAL NATURAL ENVIRONMENT: EASTERN RADNORSHIRE, POWYS, MID WALES

This example is in the middle Welsh Borderland (Figure 1.1 for location). Contrasts in natural environment can be distinguished (Figure 1.2). The wide Radnor Valley is below 250 m (800 feet) and relatively flat. It is largely enclosed by hills, the highest being the Radnor Forest range to the north-west rising to 660 m (2166 feet).

Geological underpinning for these marked landform differences is easy to interpret (Figure 1.2). The valley area coincides almost exactly with the extent of the Silurian Wenlock Shales outcrop, the least resistant rocks in the local district. Surrounding hills are formed from more resistant geology, mainly younger Silurian Ludlow Series (mudstones, siltstones and flagstones) but with outcrops of limestone and hard Precambrian rocks to the south-east.

The Radnor Valley floor is covered by silts and gravels deposited by ice and glacial meltwaters at the close of the Ice Age (*c.* 18 000 to 14 000 years ago). These give fertile well-drained soils and good water supplies to contrast with mainly thin, very acid soils in the hills.

There are appreciable variations in annual rainfall totals. In the Radnor Valley annual averages are about 1100 mm (44 inches) in the west and as low as 900 mm (36 inches) in the east. Over the Radnor Forest hills, rainfall averages as much as 1400 mm (55 inches) accompanied by greater cloudiness and lower temperatures with increased altitude.

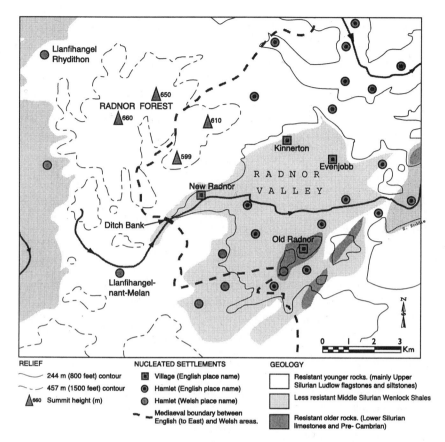

RELIEF
- ~ 244 m (800 feet) contour
- ~ ~ 457 m (1500 feet) contour
- ▲660 Summit height (m)

NUCLEATED SETTLEMENTS
- ▣ Village (English place name)
- ◉ Hamlet (English place name)
- ● Hamlet (Welsh place name)
- ━ ━ Mediaeval boundary between English (to East) and Welsh areas.

GEOLOGY
- ☐ Resistant younger rocks. (mainly Upper Silurian Ludlow flagstones and siltstones)
- ▨ Less resistant Middle Silurian Wenlock Shales
- ▰ Resistant older rocks. (Lower Silurian limestones and Pre- Cambrian)

Figure 1.2 Local natural environment and rural settlement: eastern Radnorshire, Powys, Mid-Wales.

The natural environment described in Example 1.1 is a historic border zone between England and Wales (Figure 1.2). English settlers moved into the Radnor Valley from the east, certainly by the ninth or tenth centuries, as place names indicate. By the twelfth century New Radnor was established as a small planned English strategic and administrative centre. Settlement in the valley is mainly nucleated with villages and hamlets associated with the mediaeval two-field open field system already described as typical of western regions of England.

Surrounding hill districts are populated thinly with few nucleated settlements and predominantly dispersed patterns. Place names show that these areas remained Welsh during mediaeval times. The mediaeval Lordship of Radnor, based on New Radnor Castle, was divided between 'Englishry' and 'Welshry' from the twelfth to fifteenth centuries. An old earthwork, Ditch Bank, across the upper Radnor Valley, appears to mark this ancient divide and to this day coincides with the boundary between the communities of New Radnor (English place name) and Llanfihangel-nant-Melan (Welsh place name) (Figure 1.2).

What does this local example tell us about nature and rural settlement forms? Here are some conclusions which clearly have a more widespread application irrespective of country or region:

- Flatter land with good soils, adequate rainfall and water supplies supports a denser rural settlement initially based on subsistence crop farming.
- Hilly terrain with poorer soils and too much rainfall, in the British climatic context, supports a sparse rural population whose emphasis has been on a pastoral farming economy.
- Natural forces of landforms, geology, soils and climate influence settlement forms through socio-economic and cultural variations. Radnor Valley is still characterized by English nucleated forms but adjacent areas by Welsh dispersed patterns.
- Ancient cultural influences persist for centuries in such an area but would not be influential in new countries such as the United States, Canada or Australia.

SITE-SPECIFIC SCALE

These factors will be considered when looking at siting of urban settlements shortly.

WORKPIECE 1.2

NATURE AND RURAL SETTLEMENT

Examine a reasonably large-scale map for a rural area (which could be the same map as used for Workpiece 1.1). Also refer to any information available for that area, whether in mapped or written form, regarding the geology, soils and climate.

- Attempt to relate any difference in settlement density and form to landforms. Does settlement avoid the lowest ground, highest ground or steeper slopes? Is dense settlement with more nucleated settlements mainly located on flatter terrain?
- Try to correlate landforms with soils and geology in the map area, if the information is available. Do soils and geology help to explain distributions and patterns of settlement in the area?
- Are settlement patterns in the area in any way related to water supplies, (e.g. rivers, streams, wells, springs)? Is flood hazard a deterrent to settlement in some low-lying districts?

NATURE AND URBAN SETTLEMENT

Towns and cities become established when the social and economic organization of the inhabitants requires them. Their functions as centres for trading, processing and manufacturing, administration, religious organization or political and military control are needed when society attains a certain level of development. Historically this stage seems to be linked closely with the transformation from nomadic pastoral farming or shifting cultivation economies to settled cultivation (sedentary agriculture). When this change occurs, individuals can be emancipated from full-time commitment to subsistence farming. They may become specialist traders, artisans, administrators, priests, soldiers and servants.

Archaeological evidence indicates that urban life commenced some 10 000 years ago in parts of the Middle East, as for example at Jericho in Israel. Towns existed throughout the fertile crescent of the Middle East and through South-East Asia by 5500 to 5000 years ago. By contrast, urban development came much later in north-west Europe with towns not established in southern England until just over 2000 years ago, perhaps by about 100 BC.

Each urban settlement adapts to its own unique set of natural environmental circumstances. Do these circumstances influence at different geographical scales, as with rural settlement? Consider the siting of towns and cities. An urban centre may evolve as the clear focus of a large recognizable natural region, such as Paris in its Paris Basin (regional scale). Towns may develop because of more localized situation factors, encouraging growth of urban centres in certain general situations (local scale). Furthermore natural forces have been invariably influential in the choice of the precise location of what was to become the original nucleus of a town (site-specific scale).

If natural forces offer choices regarding selection of town locations, can they also be said to affect how they develop? Is spatial form influenced significantly? Let us examine possible local and site-specific factors on town siting and urban development.

Some of the most frequently recognized of the natural environmental factors which may have encouraged the establishment of a town in a general local situation, say to within a few kilometres, include the following.

LOCAL-SCALE SITUATION FACTORS

● **Boundary zone between contrasting natural regions.** Such a situation can facilitate trading and industry because products from different areas can be exchanged, bought and sold, or processed. Examples exist all over the world as, for instance, towns situated along boundary zones between hilly and lowland regions.

● **Route focus.** Routes may be guided by the natural configuration of the land to focus on a particular local area. Gaps and valleys in hilly regions, or drier corridors through wet low-lying districts, can create such a natural focal point for roads and railways.

- **Existence of particular natural resources.** Minerals, for example, will lead to the founding of urban centres for trading and processing the extracted minerals present. Metal ores, coal, oil and salt have led to the creation of towns in many parts of the world.
- **Coastal area with a natural hinterland.** Port towns develop where there is easy access to inland regions, often because of features such as valleys or gaps, to make their local situation a natural outlet.

SITE-SPECIFIC SCALE FACTORS

Natural phenomena usually guided the selection of site for the initial nucleus of a future town at a specific location, perhaps to within a few hundred metres. Many of these natural influences can be identified by studying an actual town. Warwick, an ancient town in the English Midlands, has been chosen.

The origins of Warwick can be traced back to Anglo-Saxon times. About AD 914 a motte-and-bailey fortification was constructed – the great motte earthwork still visible and known as Ethelfleda's Mound after the Saxon princess who directed its foundation. Once the fortification was established, a civil settlement developed, no doubt benefiting from its protection and patronage. Before the Norman Conquest of England in 1066, Warwick had become sufficiently important to be designated the county town of Warwickshire, an Anglo-Saxon shire covering some 2200 km² (c. 850 square miles). Example 1.2 examines the possible natural siting factors.

Any combination of natural siting factors, both local and specific, could have influenced the original site selection for an urban centre. To those site-specific factors identified for Warwick (Example 1.2), an inland town, may be added those which encourage the siting of ports. A natural harbour or at least sheltered anchorage is an obvious near-prerequisite for a coastal port. The highest point of navigation upstream is a key factor for a river port.

Other forces need to be appraised when analysing urban siting factors. Some of those factors noted for an ancient town like Warwick are less frequently significant for towns of more recent derivation. For example, towns in 'new' countries, or towns established over the last three centuries, are less likely to have had a defensive siting as a primary consideration. In new countries a route focus, whether or not strongly guided by natural forces, is more frequently an originating influence. Some urban sites have been selected in the past for entrepreneurial, political or religious motives, with physical influences relatively less important.

EXAMPLE 1.2

URBAN SITING FACTORS – WARWICK, ENGLAND

Natural siting factors appear to have been of great importance in the location of Warwick (Figure 1.3). Local-scale factors existed. A strategic and market centre would have developed somewhere in the rich and relatively densely populated Avon Valley. Warwick also appears to have been located between two contrasting natural regions. Northwards stretched heavy clay soils, extensive forests and the predominantly pastoral farming of north Warwickshire, known as the Arden. To the south the lighter soils and gently rolling country of Feldon supported extensive arable farming. Warwick was well situated for trading with animal and timber products from the Arden and the varied agricultural products from the Feldon country.

Specific siting factors were very influential. Many occur which are evident for towns all over the world (Figure 1.3). The original fortification and later mediaeval castle were built on one of the finest defensive sites in Britain. There was a sheer 15 m (50 feet) sandstone cliff on the River Avon and land fell away in other directions to small marshy tributary valleys. A bridging point existed immediately upstream from the castle so that it could be overlooked and controlled. The location of the river crossing was pinpointed further by narrowing of the river flood plain to afford reasonably dry approaches. Construction of castle and town was facilitated by a reasonably flat site. The flat-topped hill of porous sandstone provided good water supplies from wells. A route focus existed, as the bridging point brought a south-east to north-west road to intersect with the north-east to south-west route following the Avon Valley.

Having read this section, can you now attempt to detect possible natural siting factors for any town or city with which you are familiar?

URBAN SITING FACTORS

For any town or city of your choice try to detect its possible natural original siting factors from available maps and/or your own observations and knowledge. Then attempt to answer the following questions:

● Are there any natural local-scale factors which led to the development of the town or city (e.g. route focus or boundary between contrasting regions)?

● Are there natural site-specific factors which appear to have encouraged the precise siting of the town? Can you recognize the original nucleus of the town or city? If so, what natural factors influenced the siting of that initial settlement?

Whether or not a town expands successfully must be attributable very largely to economic, social and political forces. Nevertheless, how far does its growth pattern demonstrate adaptation to natural forces? Some answers should emerge if we return to the case study of Warwick (Figure 1.3).

Following the Norman Conquest, Warwick Castle became the seat of power for one of the greatest baronial dynasties in England, the

URBAN FORM AND DEVELOPMENT

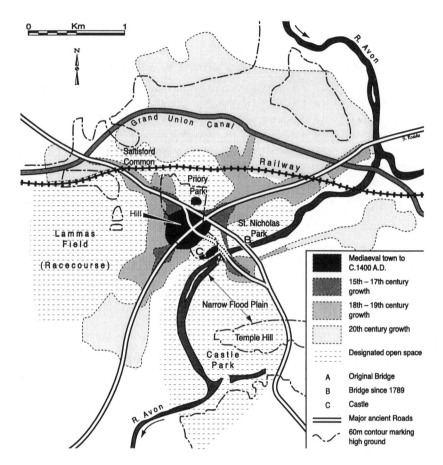

Figure 1.3 Natural features, urban siting and development: Warwick, England.

Earldom of Warwick. During the twelfth, thirteenth and fourteenth centuries the massive stone castle took shape, replacing the earlier wooden motte-and-bailey structure.

Behind the castle, on the sandstone hill, the town developed as a market and administrative centre. By the fourteenth century Warwick was a walled town. Four main axial streets ran to four town gates as part of a planned mediaeval street system. The population grew to between 2000 and 2500 people.

How did natural agencies influence the later growth and spatial form of Warwick? From the fifteenth to the seventeenth century there was limited expansion beyond the confines of the old town walls. Warwick grew towards and across the bridging point. It expanded south-westwards and north-eastwards along the valley route on elevated ground and avoided the Avon valley floor, which was prone to flooding. The flood plain remains as grazing lands and open space right through to the present day, as at St Nicholas Park and Castle Park.

Later expansion has followed the same natural guidelines; that is, along the valley sides but not across it. However, natural forces have not been the only influences on the developing urban form. The politico-economic factor of land ownership has been important in Warwick. Formerly church-owned priory lands to the north (now a public park), the Lammas Field common lands (now Warwick Racecourse) to the west, other open common lands to the north-west at Saltisford, as well as the private estate lands of the Castle, were historically not available for development. New transport links in the nineteenth century led to industrial and commercial development to the north. The construction of the Grand Union Canal was followed by that of the Great Western Railway, both major links between the Midlands and London which passed across the north of the town. Today Warwick has a population of over 20 000 and still retains its shire town functions.

What generally applicable conclusions regarding natural forces, urban siting and development can be made from the Warwick case study? Here are some of the important ones:

- Local and site-specific natural factors normally exert considerable influence on choice of location for the original urban nucleus. This is especially so where the nucleus was a rural settlement (village), as is often the case. Landforms, geology, soils and water supply are particularly significant.
- Physical environmental factors may influence the spatial form and development of towns. For example, natural features such as flood plains, steep slopes or water barriers may preclude expansion in certain directions.
- Natural factors affect choices motivated primarily by economic, social and political forces. The pace and pattern of urban growth can depend in large measure on human influences such as individual patronage, land ownership and transport facilities.

NATURAL LIMITATIONS TO URBAN DEVELOPMENT

In certain situations, nature places definite limitations to urban expansion. The example of Warwick demonstrated southward limits placed by a river and its flood plain (Figure 1.3). Other physical features which commonly affect urban form and development include steep hill slopes, lakes and of course the sea. Example 1.3 shows how the whole urban form may be influenced fundamentally by a physical barrier.

EXAMPLE 1.3

NATURAL BARRIERS AND URBAN FORM

Generalized spatial models of urban development suggest a basically concentric idealized pattern of spatial growth. Concentric zones of expansion, or rings, come to be associated with certain dominant land uses and may form recognizable functional zones. The original nucleus usually becomes the central business district and so on until the most recent outer ring of development is normally a residential commuter zone of better class, low density housing.

An instigator of this type of urban spatial model was E.W. Burgess, who analysed the form of Chicago on the shores of Lake Michigan in the 1920s [4]. The effect of this complete natural barrier has been to produce a semi-circular, rather than circular, urban form (Figure 1.4). The famous Chicago waterfront of impressive tall buildings marks where the central business district itself fronts on to the Lake Shore Drive. Today that central business zone is still defined roughly by the Loop, an encircling elevated railway system.

Other lakeside cities, such as Toronto or Geneva, and most coastal towns demonstrate similar characteristics imposed by a water barrier. Cities on large rivers or estuaries have the same constrained semi-circular growth pattern as, for example, Bordeaux (France) on the west bank of the Gironde estuary or Liverpool (UK) on the east shore of the Mersey estuary.

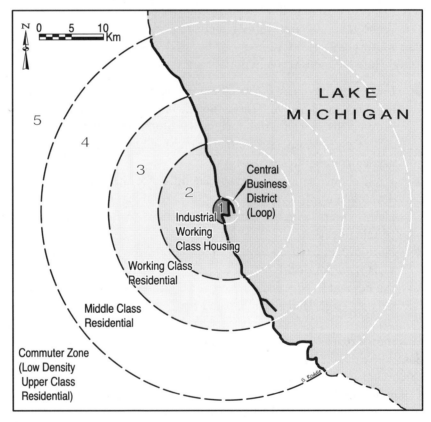

Figure 1.4 Natural barrier and model of urban form: Chicago, USA.

Particular physical environments obviously prevent a 'normal' pattern of concentric urban expansion. Some coastal urban nuclei were located on peninsulas, so producing a more linear form. San Francisco, on the peninsula between San Francisco Bay and the Pacific Ocean, has expanded southwards along the inside of the peninsula, restricted by the Bay to the east and the coastal hill ranges to the west. Valletta, the capital of Malta, was a late sixteenth century planned town located on the peninsula between Marsamxett Harbour and Grand Harbour. Expansion beyond the original walled town could not follow a normal pattern (Figure 1.5a).

Some coastal cities have developed necessarily fragmented forms determined by their island origins. These include New York (Figure 1.5b). The city originated at the southern end of Manhattan Island. Expansion northwards occurred along the island, with Broadway representing the initial axis and connecting route. Further urban growth required a series of bridges and tunnels. Links were made across the East River to the western end of Long Island where Brooklyn and Queens developed. Northwards settlement spread across the Harlem River to the mainland where the Bronx expanded rapidly. Westwards, tunnels under the Hudson link with the New Jersey mainland.

The New York conurbation, therefore, has a complex form consequent upon natural barriers. Limitations of the Manhattan Island site were a major impetus to build upwards on scarce and valuable land, culminating in the skyscraper complex of central Manhattan by the 1930s.

Many inland cities were sited deliberately in natural riverside locations which would inhibit future patterns of growth. Two such locations are confluences and meander cores. Both offer excellent defensive sites which control river crossings. The inside angle of a **confluence** provides a superb strategic site commanding bridges over both rivers immediately before they merge. Examples are the old city of Passau in Germany (Figure 1.5c) where the Inn joins the Danube, Lyon (France) on the Rhone–Saône confluence, and the American city of Pittsburgh, Pennsylvania, where the Allegheny meets the Ohio. In these locations growth has to be primarily along the interfluve and then across the rivers.

Meander cores, or land inside river loops, are protected on three sides especially where the river is incised below steep slopes or cliffs. The only land access is via the narrow meander neck. A fortification at that point protects the town and guards any bridges on both sides. In England the northern cathedral city of Durham, on the River Wear, and

17

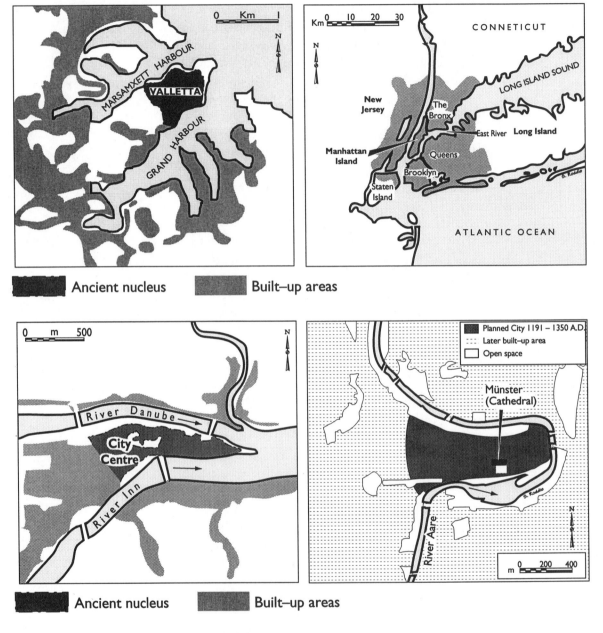

Figure 1.5 Examples of natural features and urban form. (a) Peninsula – Valletta, Malta; (b) islands and peninsula – New York, USA; (c) river confluence – Passau, Germany; (d) river meander core – Bern, Switzerland.

the town of Shrewsbury on the River Severn are in exactly that type of location. The mediaeval city centre of Bern, the Swiss capital, is similarly located, within an incised meander of the River Aare (Figure 1.5d). Urban patterns are clearly affected. The town centre development is physically restricted and expansion has to be beyond the neck of the meander or across the river.

These examples show us some clear conclusions regarding natural factors and urban forms:

- Certain natural features can influence the whole pattern of urban development.
- Physical barriers inevitably prevent the development of any symmetrical normal urban growth pattern. An urban centre may become semi-circular, fragmented or even polynuclear with several semi-independent centres.
- Other forces may disrupt an idealized concentric urban form. The Warwick example demonstrated that factors such as land ownership and transport routes have had powerful influences on the pattern of development (Figure 1.3).

So far in this chapter there have been theories and examples of how rural and urban settlements adapt to natural environment. By implication the settlements studied must have adapted successfully. They have survived and flourished. But throughout history there have been instances of inadequate adaptation because natural forces have been underestimated or ignored. Population pressures on land for farming or housing, or attraction of a valued resource, may induce settlement in naturally unsatisfactory or actually hazardous situations. Initially successful adaptation may be jeopardized by environmental changes. Example 1.4 looks at possible twentieth century inadequate adaptation by rural settlement.

What were the key factors in the national-scale disaster of the Dust Bowl in the United States? Firstly, rural settlement failed to take into account the dynamic nature of natural environment, in this case climatic variability. Secondly, there was ignorance concerning the adaptive farming methods required for wheat and corn cultivation in the drier Midwest. These same two factors are relevant today in many regions of the world. Desertification with enforced abandonment of settlements is occurring partly because of actual climatic variation but also because of inadequate adaptation through over-cropping or over-grazing.

INADEQUATE ADAPTATION TO NATURE

EXAMPLE I.4

INADEQUATE ADAPTATION TO NATURE – THE 'DUST BOWL' OF THE MIDWEST

A classic example of unsatisfactory adjustment to natural environment has been in the Dust Bowl regions of the American Midwest. During the 1930s and especially in 1931, 1933, 1934 and 1936, large areas of Oklahoma, Kansas, Nebraska and the Dakotas were victim to widespread soil erosion. Desiccated top soil became loose and was removed in disastrous dust storms. Farmland was ruined and settlements were overwhelmed. Hardship was exacerbated by economic depression and more than 25% of the farm and village people of Nebraska, Kansas and Oklahoma are estimated to have left during the 1930s.

Settlement and crop farming had spread westwards in the late nineteenth century when Midwest climates were in a wetter phase. No special adaptive farming techniques had been used; they were mainly those employed successfully in wetter regions further east. The 1930s demonstrated insecure adaptation. Exposed soil on crop lands became vulnerable to wind erosion where only 50 years previously there had a protective cover of prairie grassland.

More recently, different farming techniques have been introduced including dry farming, contour cropping and irrigation. Better land management has enabled rural settlements to withstand climatic variability in later drought years such as 1988.

Settlements in low-lying areas can be vulnerable to flood hazard. Excellent fine silty soils derived from periodic flooding make them attractive for habitation but, without careful adaptation in the form of flood defences, there is an uneasy relationship with nature. On a regional scale much of Bangladesh is vulnerable. Settlements with some 30 million inhabitants are located within 3 metres of sea-level. Severe flooding occurs both from the rivers (Ganges and Brahmaputra) during the monsoon season and from sea surges driven by tropical cyclones crossing the Bay of Bengal. Since 1970 hundreds of thousands of people have been killed and countless settlements abandoned, at least temporarily [5]. In the United States the great 1993 Mississippi floods incurred costs estimated at more than 4500 million and 6500 million US dollars for property damage and agricultural losses, respectively, with 48 persons killed [6].

At a local level, in all regions and countries, individual settlements and dwellings may not be in accord with natural processes. Many are located on low ground, such as a flood plain or reclaimed land, which is subject to flooding. Some are sited on potentially unstable slopes, because of steepness or geology, or too near the edge of receding cliffs being eroded by a river or by the sea. In mountain regions, villages may be in danger from snow avalanches at certain times of the year. Settlements in volcanic and earthquake zones are always vulnerable. In the first half of 1995, earthquakes in the western Pacific zone devastated the Japanese city of Kobe, with 5000 people killed, and the small

Russian oil town of Neftegorsk on Sakhalin Island, with some 2000 deaths.

Decisions are taken to site settlements in risky situations. Reasons include land availability and population pressure. Frequently, whether at a regional or local level, completely satisfactory adaptation to nature is not feasible due to expense or to engineering limitations.

WORKPIECE I.4

SETTLEMENTS AND NATURAL HAZARDS

Some examples have been given in this chapter of settlements not satisfactorily adapted to natural forces. Consider examples that you know locally, or from reading and filmed information.

● Describe examples where settlements are threatened or have been damaged or destroyed by slope movements (e.g. landslips, soil creep, mudflows, avalanches, etc.).

● Give examples of settlements vulnerable to flooding. Try to explain why the settlements are in such hazardous situations.

● Describe an example of a natural hazard (other than slope instability or flooding) threatening or damaging a settlement. What are the effects on the settlement?

Frequently there are efforts to control nature in order to provide more and safer settlement sites, as Example 1.5 explains.

EXAMPLE I.5

ATTEMPTS TO CONTROL AND CHANGE NATURE

From the beginning of civilization there have been attempts to control and change natural features in order to establish or extend settlements. With technological advances we are increasingly able to modify our environment.

Landforms have been altered for thousands of years to produce additional suitable building sites. Hillslopes have been terraced with retaining walls to provide more flat sites. Low-lying or reclaimed lands have been protected from flooding by the sea or rivers by constructing embankments or dykes. In the Fenlands of eastern England 500 000 hectares of land has been reclaimed from marshlands or the sea since 1640. In the Netherlands 40% of the total land area is protected land.

In most cities of the world low ground has been raised by tipping rock debris, soil or waste materials to provide a drier foundation for buildings.

The kinds of environmental modification mentioned so far occur world-wide. Sometimes there have been dramatic efforts to produce a development site for the perceived needs of a particular society. For example, in recent years a large hill has been levelled in Rio de Janeiro to provide additional flat development land. Remarkably it is estimated that over 50 000 tonnes of material was moved to construct a raised, level, defensible site for the city of Erbil in Iraq. That site has been occupied for at least 6000 years.

SETTLEMENT, BUILDINGS AND CLIMATE

The two most fundamental requirements of human beings are food and shelter, and the impact left by peoples upon areas they occupy results from activities associated with those two requirements. [7]

An earlier section showed that farming organization, partly in response to natural forces, does influence rural settlement patterns. Shelters, or dwellings, have their individual and collective impact on place and neighbourhood. They are designed very much in response to one of those natural forces: climate. Ways in which they are grouped in settlements may be also influenced by climate. Building materials depend on local natural resources, whilst landforms affect the location and configuration of settlements. Referring to rural house types, Faucher wrote: 'Each layout has been adopted in relation to the materials used, the space available and physical conditions, most important of which is climate'[8].

Human beings modify climate firstly by wearing clothes appropriate to the weather conditions experienced. This produces a body climate, or **ecoclimate**, with temperatures a little below body heat values. But clothes are not sufficient for comfort – even in the tropics, during rainfall or at night. Humans need to construct dwellings appropriate to the climate to create a comfortable interior climate, or **cryptoclimate**, with temperatures ideally maintained in the comfort zone range of 18–23°C (64–74°F). How do house types vary with climate to achieve an acceptable cryptoclimate?

Two contrasting rural house types can suggest some answers to this question. Through Roussillon, Languedoc and Provence across southern France a tall compact house type predominates (Figure 1.6a). These regions have a Mediterranean climate with mild, rainy winters and hot, dry summers. Usually three-storey houses are painted white, or a pale colour, on the exterior of thick stone walls. Light colours reflect sunshine and the walls insulate from summer heat. Relatively small windows have shutters for shade. Tiled roofs are flat or gently pitching in a climate where snow is exceptional and rainfall only seasonal. Traditionally the ground floor is used for animals, farming implements and storage of bulk commodities such as wine. Living accommodation is on the first floor and bedrooms on the second floor. The origins of this type were as farming village houses, often in tightly nucleated settlements with narrow, shaded streets on either a hilltop or valley side site in typical Mediterranean fashion, initially for defence.

Climatic conditions are very different in the Alps. Reasonably warm summers have considerable rainfall and cold winters produce

Figure 1.6 Climate and house design. (a) Tall compact house type (Provence, France) – Mediterranean climate (warm temperate, summer drought, winter rain). (b) Chalet house type (Tyrol, Austria) – continental climate (cool temperate, summer rain, winter snowfall).

heavy snowfall. Although house types vary locally throughout the French, Swiss, Austrian and German Alps they have many common features and the easily recognized chalet architecture has evolved (Figure 1.6b). How is it adapted to climate?

Alpine houses are of strong timber construction with steeply pitched roofs and wide eaves. Windows and doorways are therefore well protected and normally there are sheltered inset balconies. The wide eaves give protection from rain and snow whilst the pitched overhang ensures that any roof slides of accumulated snow will fall away from entrances and windows. The three storeys are used in much the same way as in southern France except that fuel storage is more important on the ground floor. First-floor living areas open out on to the sheltered balcony which is usable in any weather conditions.

Alpine villages are usually located on valley floors or valley side ledges – the available flat sites. Aspect is important in mountains so that preferred sites are on the sunny (south-facing) sides of valleys rather than the shady sides. The German terms *sonnenseite* and *schattenseite* describe these contrasting situations. Settlements tend to an open form: dwellings are well spaced, with intervening grazing areas for cattle. Farming is based on dairying and the cattle are brought down from the high mountain pastures from September to April.

This comparison of two traditional rural house types, which use their storeys for precisely similar functions, demonstrates that:

● house design is influenced by climate;
● house design is affected by the natural building materials available;
● grouping of dwellings, imparting the essential sense of place to a village, reflects both climatic and socio-economic forces such as farming organization and defence.

This section has looked at traditional house types with origins going back many centuries. However, the impact of climate is still significant for all buildings and settlements. Through modern technology acceptable cryptoclimates can be maintained with central heating and air conditioning, but costs are involved. Planners and architects have to consider microclimatic effects on modern urban developments such as housing estates, shopping precincts or business complexes. Aspect and orientation to prevailing wind directions are especially important.

WORKPIECE 1.5

BUILDINGS AND CLIMATE

Consider the traditional house types in your home country or region. These are more likely to be obvious in rural areas but may exist in urban centres also.

● Describe the features of the houses which appear to be responses to local climate conditions, e.g. roof shape, window size and shape, overhang for shelter, etc.

● Discuss how far house design is a response to the availability or non-availability of local building materials, e.g. stone, timber, clay for bricks, suitable vegetation for roofing, etc.

● Draw a sketch of a typical house, labelling features influenced by natural environmental factors (climate, building materials, etc.).

NATURE, SETTLEMENTS AND SUSTAINABILITY

This chapter has touched on aspects of the interactions between settlements and natural environment. Siting, form and development of settlements are influenced by nature. Not all settlements are adapted satisfactorily to natural forces. Building design is to some degree responsive to climate. Therefore nature influences settlement, but what impacts do settlements have on nature? Can settlements live in harmony with nature and to what extent are they mutually sustainable? This final section looks at answers to those key contemporary questions.

Today we live in a world of accelerating population growth, urbanization and industrialization. Large conurbations have increasing effects on environment. All economies at some stage were based on subsistence agriculture. In such societies village and nature were mutually sustainable. The Brundtland Commission (1987) defined sustainable development as 'development that meets the needs of the present without compromising the ability of future generations to meet their own needs'[9]. In subsistence economies local farming provides practically all food and raw materials for the community. Waste products are returned to the land to maintain fertility. Settlement and nature exist in dynamic equilibrium with land and environmental quality sustained for future generations.

EXAMPLE 1.6

THE MODERN CITY AND NATURE

The modern industrialized city consumes enormous quantities of natural resources as food, water, energy and raw materials. These can be regarded as inputs to the urban resource utilization system (Figure 1.7). From this system huge outputs of waste products and pollutants impact on the natural environment. Solid wastes are tipped on landfill sites as the easiest form of disposal. Coastal cities may deposit solid wastes at sea. Sewage, treated or untreated, is also disposed of on land or into water.

Fossil fuel burning produces some solid wastes – for example coal ash from power stations – and gaseous emissions into the atmosphere, with global repercussions. Carbon dioxide is responsible for about 50% of the enhanced greenhouse effect which many scientists believe is causing global warming. Sulphur dioxide and nitrogen oxides cause acid rain, with damage to forests and freshwater ecosystems. Concentrated urban pollution results from fossil fuels used in motor vehicles and industry and comprises a cocktail of gases potentially hazardous to human health. Chlorofluorocarbons (CFCs) are produced for specific manufacturing applications and are considered to be the chief cause of the weakened stratospheric ozone layer. This allows more dangerous ultraviolet solar rays to reach the earth's surface. Inevitably these outputs have adverse effects on all natural systems to the detriment of sustainability.

By contrast modern cities are vast resource users with enormous outputs of waste materials (Example 1.6). Urban areas need to reduce their polluting outputs by utilizing resources more economically, for instance through energy conservation. There should be more recycling back into the system (Figure 1.7) with better organization of categorized waste (e.g. paper, wood, glass, metals). Reprocessing of waste for fuel, fertilizers or building materials should be introduced more widely. Refuse disposal should be controlled so that both surface and ground water are protected. Less polluting and more efficient urban transport systems are essential.

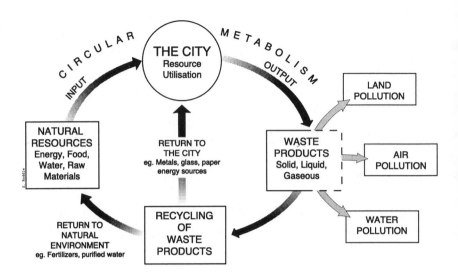

Figure 1.7 The modern city and nature: a resource system.

More used resources should be returned to the system. Some can be returned directly to the city, such as reprocessed raw materials or purified water. Some can be recycled to the natural environment to replenish resources as, for example, sewage sludge treated organically by reed filter beds and used as fertilizer for farmland. In this way a circular metabolism can be established for the city which gives back to nature as much as it takes out.

Cities and nature are only mutually sustainable if they function in harmony. Should they be more spread out or more compact? Can an aesthetically pleasing built environment coexisting with attractive diverse natural features be achieved? Should early twentieth century ideals of well-designed buildings and communal open spaces be reinstated? Modern electronic technology and less interrelated heavy industry mean that there is less rationale for large continuous urban areas. Urban designers suggest that such areas could become polynuclear: an interconnected network of local centres separated by landscaped open areas. Concepts of neighbourhood could be revived as people would live, work and shop in the same locality. Could private car journeys be reduced? How will the settlement pattern we adopt affect transport needs and the quality of life?

Some local experiments are attempting to restore a sustainable relationship between settlements and nature. A housing cooperative at Kiel (Germany) has the stated social aim of creating 'an environment suited to the concept of the community and the ecological realm of the holistic environment beyond the individual home'[10]. The development for 21 houses has its own on-site energy generation and waste disposal.

On a city-wide scale, Barcelona (Spain) has undertaken schemes connected with hosting the 1992 Olympic Games. The overall plan for the inner city, streets, improved public transport, 150 interlinked parks and public places, linear coastal park and new waterside district is a blueprint for the sustainable city in closer liaison with nature.

Such mutual sustainability requires long-term policy and financial commitment. There needs to be an integrated approach at local level including, for example, strategies for cleaner, efficient public transport to reduce congestion and pollution. Local strategies should fit international environmental agreements like those on biodiversity and greenhouse gas emissions at the Rio de Janeiro Earth Summit in 1992. Public participation is vital at city, neighbourhood and individual site levels through meetings, discussions and exhibitions.

In answer to the two questions raised at the beginning of this section it can be concluded, firstly, that modern cities have tremendous adverse impacts on nature. Secondly, and more optimistically, it can be said that settlements can coexist with natural systems for mutual sustainability. Policies should aim to restore a dynamic equilibrium with nature. Could the 'Garden City' ideal of Ebenezer Howard provide a way forward, or are there other directions which could be explored? As Sir Richard Rogers concluded in his final Reith Lecture for 1995, *Cities and Sustainability*: 'Sustainability should be the ethic of the development process and, indeed, the ethic of our age'[11].

SUMMARY

Nature has considerable influence on rural and urban settlement forms. However, natural forces are not the only influences. Rural settlement patterns have evolved through adaptations of human societies to those natural forces according to their extremely varied socio-economic organization. Natural environmental forces offer possibilities and limitations for urban siting and development which are interpreted through human choice. In certain localities natural factors may have dominant influences on urban form as physical barriers to expansion or as natural breaks to fragment urban areas.

Settlements have not always adapted perfectly to nature and may be vulnerable to natural hazards. Conversely the natural environment may be modified to provide more secure and comfortable settlement siting. Natural factors affect design, disposition and materials used for buildings to impart essential physical and aesthetic character to settlements.

Settlements can have adverse effects on nature. Cities can produce congested, polluted and unattractive environments. Reduced resource usage, more recycling and urban replanning are required to restore a more harmonious and sustainable coexistence between settlements and nature.

Nature influences settlements, settlements have impacts on nature. This complex interrelationship is fundamental to the understanding of many aspects and issues concerning buildings, neighbourhoods and places to be discussed in the following chapters of this book.

CHECKLIST

Having read Chapter 1 there are several important points that you have learnt regarding the relationships between nature and settlement:

● Natural forces influence the siting, patterns and density of rural settlements.
● Human factors, especially cultural traditions and agricultural organization, influence rural settlement types through selective adaptation to the natural environment.
● Natural factors affect choice of urban settlement sites and the spatial development of towns and cities.
● Some settlements are not adapted adequately to nature and are susceptible to natural hazards, whilst the growth of others may have been assisted by deliberate modifications of the environment.
● Natural forces influence design, positioning and building materials of dwellings, especially traditional regional house types.
● Settlements have increasingly adverse impacts on nature. We need to pay more attention to urban planning and design, reduced resource usage and recycling so that more aesthetically pleasing cities can coexist with nature for mutual sustainability.

These points will give you valuable background to the aspects of buildings, neighbourhoods and places discussed in the following chapters.

REFERENCES

1. Thorpe, H. (1964) Rural settlement in the British Isles, in *The British Isles: A Systematic Geography* (eds J.W. Watson and J.B. Sissons), Nelson, London, pp. 358–379.
2. Thorpe, H. (1964) Rural settlement in the British Isles, in *The British Isles: A Systematic Geography* (eds J.W. Watson and J.B. Sissons), Nelson, London, p. 360.
3. Roberts, B.K. (1977) *Rural Settlement in Britain*, Dawson, Folkestone, p. 19.

4. Park, R.E. and Burgess, E.W. (1925) *The City*, University of Chicago Press.
5. Houghton, J.E. (1994) *Global Warming: The Complete Briefing*, Lion Publishing, Oxford, p.93.
6. Wharton, G. (1995) Managing river environments: the way forward, *Geographical Magazine* LXVCII 6 June, pp. 53–55.
7. Finch, V.C. and Trewartha, G.T. (1949) *Elements of Geography*, 3rd edn, McGraw-Hill, New York, p. 543.
8. Faucher, D. (1969) quoted in Pinchemel, P. *France: A Geographical Survey*, Bell, London.
9. World Commission on Environment and Development (1987) *Our Common Future*, Oxford University Press, Oxford.
10. Neil, B. (1995) Pioneers on the Urban Frontier, *Planning Week* **3**(16) 20 April, p.13.
11. Rogers, Sir R. (1995) *Cities and Sustainability*, BBC Reith Lecture Series, broadcast BBC Radio 4, 12 March 1995.

FURTHER READING

Daniel, P. and Hopkinson, H. (1979) *The Geography of Settlement*, Oliver and Boyd, Edinburgh.
Hoskins, W.G. (1988) *The Making of the English Landscape* (additions by C. Taylor), Hodder and Stoughton, London.
Owen, S. (1991) *Planning Settlements Naturally*, Packard Publishing, Chichester.
Short, J.R. (1991) *An Introduction to Urban Geography*, 2nd edn, Routledge and Kegan Paul, London.

SETTLEMENTS AND GROWTH

PETER LARKHAM

THEME

How do places grow? Why do they sometimes decline?
The theme of this chapter is the growth of settlements, from
villages to great cities. The focus of attention is on the
changing settlement form from early origins through development and growth, and even into decline. What forces are
at work? How do the activities of different interests interact
in the course of change? The processes and actors shaping
form are studied, and a wide variety of social, economic and
political forces are illustrated. The case studies deal with
both large-scale form (whole settlements) and with very
small-scale changes (individual plots and buildings). The
basic ideas contained in this chapter will allow analysis of the
form and development of almost any settlement. They provide a vital foundation for understanding and intervention.

OBJECTIVES

After reading this chapter you should be able to:

● understand better the ways in which settlements develop and
   change;

● recognize the physical characteristics of growth and decline;

● examine the forms of settlements and suggest why there are
   differences;

● understand the complexity of the processes and products of
   settlement change;

● identify the diversity of actors in the processes of change.

INTRODUCTION

A general understanding of the way in which today's settlements, both
urban and rural, have come about is fundamental to understanding the

way in which they work socially, economically and politically. Increasingly over the last two decades, the built environment professions have come to recognize the need for this knowledge of historical and current form and process. Indeed, in some cases this detailed understanding is helping to shape the form of new developments. Individual buildings are being designed with ever greater respect for the context of surrounding street and building form [1]. Even large extensions to existing towns are using historical design precedents, including Poundbury, the extension to Dorchester being developed by the Duchy of Cornwall.

Although settlement forms themselves are usually relatively simple to identify and describe, the forces shaping those forms are many and varied and they are interlinked in complex ways. Different processes can sometimes result in similar forms on the ground. Understanding and untangling this complexity demands a keen attention to detail.

There are other reasons to support a general concern for the physical shape of any settlement.

> The first impact a town makes is through its plan and the remembered way of finding one's way around. ... along with layout comes an awareness of urban scenery – the townscape – for in reality the two-dimensional map is translated into the three-dimensional array of buildings. Moreover, those buildings reflect the dates when they were built, so there is also the fourth dimension of time. Elements of the townscape are used to find our way around.[2]

Some of these reasons will be examined in later chapters of this book. The first section of this chapter discusses ideas and ways of looking at towns. The second examines some of the factors shaping settlement form and gives examples of their operation at both large and small scales.

Although this chapter mentions numerous examples, excellent advice was given by Jane Jacobs in 1961:

> The schemes that illustrate this book are all around us. For illustrations, please look closely at real cities. While you are looking, you might as well also listen, linger, and think about what you see.[3]

The study of the physical form of settlement is commonly known as **urban morphology** (although both urban and rural settlements can be analysed in the same way). Our knowledge of developing urban form comes from a wide range of disciplines, including planning, architecture, geography, history and archaeology. Each discipline has contributed

MORPHOLOGICAL IDEAS

meticulously researched examples together with ideas for unifying principles of form and its development.

HOW CAN WE MAKE SENSE OF COMPLEX FORMS?

All settlements – even those which look amorphous (shapeless) or unplanned – can be broken down into several key elements. The geographer M.R.G. Conzen has suggested that the street pattern, plot pattern and building structures are most important [4]. The network of streets is laid out first, and tends to persist longest through history. Streets define street-blocks, within which there is a pattern of property (or plot) boundaries. As property is bought and sold through the years, these boundaries may change, and they do so much more rapidly than the street pattern. But still, in many modern European towns, there is clear evidence for plot layouts persisting from the thirteenth to fifteenth centuries: the mediaeval period. Within each plot there is usually one or more buildings. But buildings decay and are changed much more rapidly than plots – on very few of those mediaeval plots is there still a mediaeval building! So, in terms of longevity, there is a hierarchy of streets, plots and buildings in that order. In the modern urban landscape, change tends to occur faster and on a larger scale – there could be several generations of buildings on the same site during one century – but Conzen's concept is still applicable.

After any series of plots is laid out, its owners (or tenants or occupiers) will inevitably start to make changes. Large plots may be subdivided, with the original owner keeping one part and selling or leasing the other. Several plots may be amalgamated, if a developer wishes to construct a large development; and, of course, this development would remove most of the evidence of earlier plots and buildings. So we can recognize a progression of change in plot patterns, and it is likely that the oldest plots will have undergone the most severe changes from their original state, while more recent plot layouts are more immediately recognizable.

There is also a recognizable progression, or cycle, of buildings on a plot. Generally, the first building will be on the street frontage of the plot: this will be the most noticeable and imposing building, sometimes called the **plot dominant**. Through time, the uses of the plot and building will change, and there will be pressures to extend the building either upwards or across the rear of the plot. This may take the form of a completely new but much larger building. Over time, more and more of the previously open plot is covered by buildings. Sometimes, even 100% of the plot is covered: this is the high point, or **climax phase**, of the cycle.

After this may come a partial or complete clearance before total redevelopment, when the cycle would begin again. This cycle (Figure 2.1) was first recognized by Conzen for the typical British mediaeval plot, the long but narrow **burgage** (named after the landholders in mediaeval towns, the **burgesses**); but it has also been shown for nineteenth century industrial towns and twentieth century suburbs.

Plot/burgage series

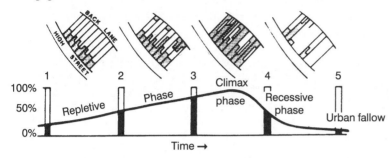

Figure 2.1 The burgage cycle (redrawn after M.R.G. Conzen).

These concepts from urban morphology are best demonstrated by example. T.R. Slater has carried out a number of detailed morphological analyses of mediaeval English towns based on Conzen's ideas. They have refined the idea of the **plan unit**, using old maps to identify regularities in the characteristics of the town plan, most notably in lengths and breadths of plots. This may suggest different growth phases for a town – as plots laid out in a planned extension of the thirteenth century are likely to have distinctly different dimensions from those of the fifteenth century, for example. More evidence from archaeology and historical research might fix absolute dates for these phases. Slater has clearly shown that the majority of English mediaeval towns are of composite nature. They consist of a number of distinctive plan units, each reflecting the particular circumstances of their creator and period of creation. Even towns such as Ludlow and Lichfield, which historians had considered to date from one phase of planning, can be shown to be much more complex[5].

This plan analysis is shown in Example 2.1 and Figure 2.2 for another small mediaeval town, Bewdley. Workpiece 2.1 develops the skills of recognizing detailed patterns of town growth.

MAKING SENSE OF A
SPECIFIC PLACE

WORKPIECE 2.1

SEEING PATTERNS IN URBAN FORM

You will need a pre-1940 large-scale map (1:2500 is best) of any older town centre (not New Towns). Local libraries or archives should have these. Cover the central part of the map with a sheet of tracing paper and, using a soft pencil, identify:

- the street network (including small alleys);
- any town walls or castle (look for curving streets which might be built next to old fortifications; look also for evidence of place names such as Castle Street or Northgate);
- the oldest church (usually on the highest point of the town);
- particularly wide streets in the older part of the town (these may be market places and might have associated names, e.g. Horsefair);
- buildings in the middle of streets, without any associated open plots (these might be market halls, or shops built on the site of open street market stalls, and have infilled a wide market street).

Having identified these important features, try to see any patterns in plot shapes or sizes: groups of narrow and long, or short and fat. You might find other types of plan units; for example, a formal Georgian square of houses surrounding a church on the edge of the older town.

These are first steps to making a plan analysis and to understanding some of the patterns of growth in a town's structure.

THE SURVEYOR'S IDEAL AND THE REALITY OF THE TOWN

Many idealized town plans are known, from all periods of history. Some were clearly intellectual design exercises, never to be built, such as some of the concentric plans of the Italian Renaissance. Others were built, but it is clear that there were often problems with implementing the ideal of the surveyor/designer. It is also true that places once thought of as 'organic', or unplanned, can demonstrate traces of regularity and planning when examined closely.

In the mediaeval period, it was very unusual for complete towns to be created anew on a cleared site. Many were, as we have seen, large planned extensions grafted on to an existing pre-urban nucleus – possibly a village or castle. Existing plots were rarely interfered with in this process, as any centralized power structure to permit this wholesale change rarely existed. The town of Wolverhampton, for example, was shared between three manorial landholders and developed by them in separate ways and directions. Many villages, particularly in north-east England, also have very regularly laid out plots.

There is much debate as to how far this is a result of deliberate planning, and at what date. One idea is that the great mediaeval landholders regularly planned both urban and rural estates to increase income from rents and to simplify its collection.

Even where developments were large, the ideal planned layout and the reality of development often differ, as in Ludlow. Here the original

EXAMPLE 2.I

A PLAN ANALYSIS OF MEDIAEVAL BEWDLEY

Bewdley is a small town in the English midlands, established where a prehistoric trackway crossed the River Severn. A royal hunting lodge and enclosed deer park were built here but the name, from the French *beau lieu*, is not recorded before 1215. Burgages, a legal form of land tenure, are not recorded before 1367. The Countess of March's new town here was on the hillside between the deer-park and the Forest of Wyre. Increasing river traffic, and a new bridge built in 1447, moved the focus of the settlement to the riverside.

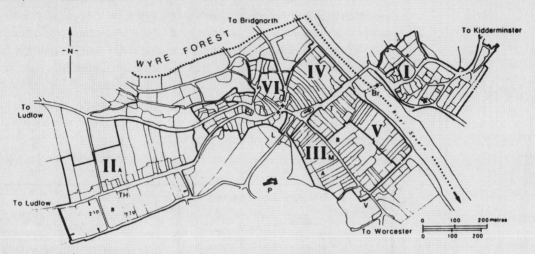

Figure 2.2 Plan analysis of Bewdley (T.R. Slater, 1990; reproduced with permission).

Slater's plan analysis [5] identifies six plan units, some in two subsections. The oldest part of the town is unit I, Wribbenhall, on the east bank and known from 'Domesday Book' (1086). Its plots are irregular. Unit II is the Countess's planned town, IIB being the regularly laid-out burgages with a rear service lane, and IIA being developed on fields. Unit III is a clearly planned street parallel to the river with regular plots on its western side. The northern side of the wide market street leading to the bridge has particularly long, narrow and curved plots (Unit IV). The south side of this road, and the riverside, form Unit V. Unit VI contains irregular plots around the curving approaches of three roads to the Welch Gate, which was known by 1472.

The plan analysis emphasizes the different characteristics of each part of the town. Documentary evidence gives some dates for some key features. What emerges is a series of phases of growth, of different characters – some planned and regular, some unplanned – but Unit II clearly represents something different: a planned layout which was quickly overtaken by events as the riverside area developed much faster.

It is interesting to see how the mediaeval surveyors designed new urban areas, and how they related to existing planned and unplanned districts. Unit II is physically separated from Unit I, the oldest part of the town, by the river and hillside. Unit III, also planned, connects to the main street but with minimal disruption, as it expanded out to the south.

The pattern is complex. It is the close fusion of planned and unplanned, regular and irregular elements that led to the older idea that the great majority of towns grew by an organic, piecemeal, unplanned process. Morphological plan analysis shows that this is often wrong and that there are many small phases of planning activity.

layout was obviously thought out carefully (broad main streets, narrow network of back alleys) with regularly proportioned plots, but it has since been developed with considerable irregularities on a steeply sloping hillside. A very active market in land at an early period soon destroyed the planned regularity of plots, so here the amalgamation and subdivision processes described earlier can be seen at work.

Even in the twentieth century, ideals and reality often diverge. The post-war plan for Worcester suggested comprehensive clearance, restructuring and rebuilding. Although some mediaeval street alignments would be kept, the streets would be straightened and widened; older buildings fronting on to them would be redeveloped on newly laid-out plots. In reality, this did not happen. Individual buildings have been redeveloped, and the creation of three large shopping centres has required considerable demolition and plot amalgamation; but still the town retains a recognizable mediaeval structure with predominantly Georgian and Victorian buildings[6].

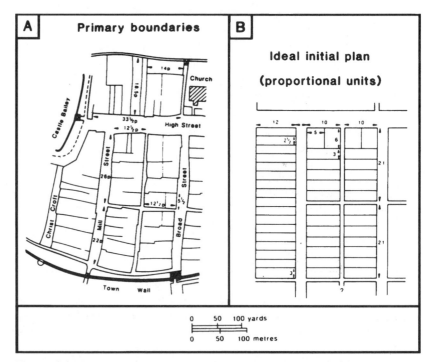

Figure 2.3 Ideal and reality in mediaeval Ludlow (adapted from T.R. Slater, 1990; reproduced with permission).

Why should land use be considered when studying changing settlement form? Patterns of land use change even more rapidly than individual building structures. Land uses are essentially temporary, but they have profound effects on form. The planned land use for an area, whether in a Roman, mediaeval or modern town, will to a large extent determine the form of new streets, plots and buildings. As a settlement grows, land uses will change – new uses coming in and old uses moving to other areas. The incoming uses may well lead to the creation of new streets or plots or buildings through redevelopment. The displaced old uses, however, are more likely to locate in older areas and to adapt them, rather than redevelop them. As uses change so buildings change, for example from a high-quality single-family house with servants' quarters, to offices and then to student bedsits. Thus some urban forms, long lasting as has been suggested, survive because of their ability to adapt to new uses. Many of the late mediaeval timber-framed houses in Ludlow were given Georgian brick fronts to improve their appearance, and now are used by antiques retailers. There has been a clear movement of the core business district of central Newcastle-upon-Tyne away from the riverside, spurred by the enormous Eldon Square shopping development. The great Georgian classical-styled buildings have become occupied by low-rent fringe uses, and have suffered from inadequate maintenance. Some are now being refurbished and their upper floors converted to residential use.

This section has introduced some ideas and tools for understanding and describing settlements and their patterns of growth (or decline). Patterns of streets, plots and buildings can be most informative. The next section deals with a fundamental question for every settlement, or indeed for every part of every settlement.

THE IMPORTANCE OF LAND USE

A number of significant factors outside the settlement itself influence changes in settlement shape, size and structure, since any settlement acts as a focus to a wider hinterland with which it is in constant interaction. Gottmann has suggested that there are several adaptive forces in society which, directly or indirectly, influence the life and shape of urban areas: these are demographic, economic, technological and cultural forces. He considers that cultural influences are the most obvious and best-established factor, but also the most difficult to assess [7]. The next sections deal with several key factors which, taken together, encompass all four of Gottmann's suggestions. They operate at all scales

WHAT FORCES SHAPE URBAN CHANGE?

in all settlements and throughout history. The problem in discussing each separately is that, in real life, most of these factors are complexly interlinked in a web of causes and effects.

POPULATION GROWTH OR DECLINE: A DEMOGRAPHIC FORCE

Population fluctuations have a most significant impact on settlement form. The most severe impact is the destruction of a settlement by abandonment. Western Europe, and the UK in particular, has a wealth of such settlements – 'deserted villages', which folklore suggests were depopulated owing to the Black Death from 1349. Although the Black Death did kill between 25 and 50% of the UK population in its several outbreaks, documentary and archaeological studies show that many depopulations occurred gradually, culminating in the widespread clearance of land by landowners in the 1700s to graze sheep.

Throughout the industrial period in western society there has been a major move of population from rural districts to urban areas where (it was felt) industry, and therefore jobs and income, would be found. The industrial cities grew enormously, swallowing formerly independent neighbouring districts and villages in the process. In May 1911, for example, Birmingham tripled in size by annexing over 30 000 acres (12 140 ha) of Staffordshire, Warwickshire and Worcestershire. This scale of growth led to problems of suburban sprawl, most clearly evident in the United States. In the UK the supply of land is less, and there is significant competition for agricultural and leisure uses. The population of the UK was more than 90% urban by 1990.

There was concern in the 1930s about the continued sprawl, particularly around London, spurred by easy commuting on the new suburban railway lines. In the post-war period, the concept of the **Green Belt** was developed and implemented around several major conurbations and some particularly vulnerable towns, such as Oxford. Green Belts have had some success in containing urban sprawl, but development has tended to 'leap-frog' the designated belts to surrounding towns and villages, particularly where there is easy access to the main city for commuting. Such places have become little more than dormitory suburbs.

Both the UK and United States have shown trends of population movement away from the central cities to peripheral areas. In extreme cases, such as Cleveland, Ohio, this has led to '**doughnut**' cities, with houses and land at the centre simply being abandoned and lying derelict. In the UK, this trend was reinforced in the 1950s and 1960s by slum clearance programmes, which razed large areas of inner urban high-density terrace slums. Not all of the local population could be rehoused in the replacement high-rise flats, and so large new municipal estates were constructed at the urban fringe to rehouse the displaced populations.

In the Third World, the urban immigration occurred more recently but with even more significant effects. Urban infrastructures could not cope with the vast population increases, which led to the rise of 'multi-million' cities, and the majority of the migrants live in appalling conditions in peripheral squatter slums.

WORKPIECE 2.2

THE INFLUENCE OF TRANSPORT

You will need a sheet map of about 1:10 000 scale of an urban area. Cover it with a sheet of tracing paper and identify, using different colours:

● canals or navigable rivers (note any dock areas);
● railways (note any goods sidings and stations);
● main roads;
● airfields;
● motorways.

Note how canals and railways penetrate into the heart of the built-up area. Canals, usually built in the 1700s, often came before the settlement, which spread out around them. Railways cut into the heart of settlements, often through the slum areas (where land was cheap). Note how both canals and railways are currently neglected, e.g. open spaces where former wharves and sidings stood, disused stations, etc.

Note the dense road network, and consider the powerful impact of the private motor vehicle on modern urban form.

The city centre, with its perceived concentration of job opportunities, is the preferred location ... but that area is incapable of assimilating the numbers. People in the lowest social groups are displaced outwards just as many of the new immigrants move directly to these squatter settlements at the margin. The term squatter has a specific connotation for it implies that the residents have no legal right to the land ... the only land available to them is waste or unwanted land which most often is unused because of environmental problems such as the extremes of flooding or water shortage[2].

EXAMPLE 2.2

THE THIRD WORLD CITY

The majority of Third World cities are caught between the desires to modernize and urbanize, thus participating in all the perceived benefits of urban life in the developed world, and the major problems which such processes develop. Some countries, such as Cuba, have deliberately resisted this trend, since they see such urbanization as inevitably bringing major socio-cultural and socio-economic change – that is, westernization.

In recent decades, though, the population migrations to Third World cities have created immense pressures on land. These immigrants are from rural areas, are generally unskilled for most urban employment, and have often moved because they are extremely poor. They are therefore not well placed to succeed in the cities. They cannot afford to live in an urbanized manner, purchasing or renting property, and so they colonize poor, marginal land. This is often, but not always, on the urban fringe. In the case of Manilla (the Philippines), the major squatter settlement known as Tondo is in a central location, close to the harbour and the Pasig River: it occupies land devastated during the war. There are many other such settlements closer to Manilla's fringes.

The key problem with the squatter settlements is that, as they are illegal and were unplanned, they have few, if any, facilities – drainage, sewerage, water or power supplies, or roads and other transport facilities. Throughout much of the Third World, attempts have been made to provide the most basic of these facilities. Municipal authorities realize the financial impossibility of bringing all of the squatter housing up to even basic levels: by improving services, they condone the illegal squatting, and encourage the squatters to improve their own dwellings. By this gradual process, the squatters should be brought into the urban environment, economy and society.

In Tondo, however, where the World Bank provided finance for such improvements in the 1970s, this has perhaps been too successful. A thriving property market has developed and rents have risen – perhaps because of the area's centrality and thus convenience – but this means that the poorest are again being squeezed out [2].

Even where upgrading is successful, the physical characteristics of the legalized squatter districts are far different from any other parts of the city. They are of much higher density; housing structures are uncontrolled; street and plot patterns are much less regular. Even if legalized, they are likely to present environmental problems for the future.

THE ECONOMY AND SETTLEMENT FORM

The economy plays a vital role in determining form. In particular, fluctuations in global, national and local economies have great influence on changes in urban form. A variety of economic cycles of boom and bust have been shown to exist; long-term oscillations with a wavelength of 15–20 years are common. These are evident as far back as records exist, but at present there are also many short-term fluctuations which may be the result of political decisions (such as the sudden UK house price rise in the late 1980s). In particular, cycles can be seen in various measures of construction activity, including building permits, brick and tile production, and building value. There are also cycles in various types of construction, including housebuilding and office development.

The timing of cycles differs across the world with, for example, North America and Europe being at opposite ends of the cycle in the nineteenth century. There may be local differences, particularly when a local industry becomes so prosperous that it can withstand even a national decline; this was the case with the South Wales coalfield area at the turn of this century [8]. Generally, in times of prosperity, the amount of construction rises rapidly. This rise may be so rapid that over-supply of one particular building type may occur, leading to an equally rapid decline, as happened in the office sectors of several major western cities in the late 1980s. In times of prosperity, rates of household formation grow, and housebuilding increases to cope with this demand. Thus economic cycles are closely linked with a variety of other cyclic trends: 'economic cycles impinge ... on population cycles, which in turn affect the economy'[9].

One of the main features of the urban landscape that reflect these cyclic factors is the **fringe belt**. A succession of ring-like areas, quite different in physical form from surrounding parts of a town, was first recognized in a study of Berlin in 1936 (Figure 2.4).

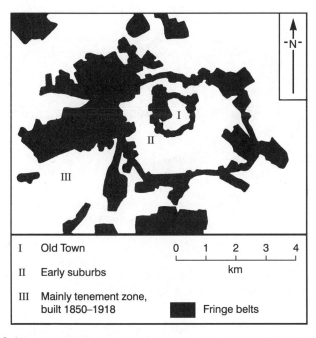

I	Old Town	
II	Early suburbs	
III	Mainly tenement zone, built 1850–1918	

0   1   2   3   4
km

■ Fringe belts

Figure 2.4 Berlin's fringe belts as seen by Louis in 1936 (J.W.R. Whitehand, 1988; reproduced with permission).

It is clear that settlements do not steadily expand outwards, but urban growth has phases of acceleration, deceleration and standstill. At times of little or no growth – often economic slumps – land values are low. Land-users needing large sites can buy them cheaply at the edge of the settlement, forming a belt of less dense development. This forms a barrier to further expansion in the next period of prosperity and growth, so development leap-frogs the belt and another ring of more dense development is formed. The repetition of phases of stagnation and outward growth creates a roughly concentric arrangement in which predominantly dense residential districts alternate with fringe belts of mixed low-density uses – including hospitals, parks, schools, colleges and universities, and sports facilities[10].

In the UK, this pattern of growth can be generalized to form a simple model, bringing together the annular fringe belts, the typical dates and types of architecture built in the boom periods, and the changes in transportation which facilitated the outward spread (Figure 2.5).

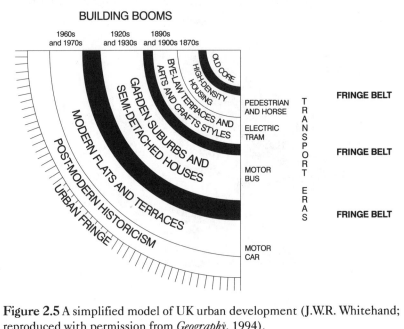

Figure 2.5 A simplified model of UK urban development (J.W.R. Whitehand; reproduced with permission from *Geography*, 1994).

Another view of the complex interaction between economics and settlement form is that of the urban sociologist Manuel Castells. This is a more theoretical perspective. For Castells, the form of a town is a direct product of a specific social structure [11]. Within this structure the economic system can be analysed in terms of four factors:

● **production**: functions producing goods and information, e.g. industry and services;

- **consumption**: functions associated with the individual, social and collective use of these products, e.g. housing;
- **exchange**: between or within both production and consumption, e.g. transport and communications;
- **administration**: reflecting the relations between production, consumption and exchange, e.g. planning and land-use policy.

The nature of the social system – such as capitalism in much of the developed west – determines which of these factors is dominant. In capitalist countries, he argues, the economic system is dominant and the production factor plays a large part in shaping urban form – creating, for example, the '**company town**'.

Many classifications of settlement function have been made. Although most are rather simplistic, it is evident that, in many cases, function has greatly influenced form at some point in history. Defence, for example, has led to towns growing up around earlier castles, as seems to have happened in Ludlow; later, these towns have needed defensive walls themselves. In continental Europe, where warfare was significant until relatively recently, many towns received large 'arrowhead' gun bastions and wide moats from the 1600s. One of the best-preserved examples is Valletta (Malta). Such fortifications constrained outward growth for a long time. In those places where they became redundant and were demolished, large areas were opened up for development: typically, wide tree-lined streets, parks and public buildings were built, as in Vienna. (In fact the word 'boulevard' has the same root as 'bulwark'.)

Most towns, and even some villages, have some communal functions which use large public open spaces. The village green common in parts of England and north-central Europe is one such space; widened streets and marketplaces in towns are another. But, as functions change over time, some of these marketplaces have become infilled with later buildings – including town halls, market halls and shops.

From the industrial period, many places have become identified with particular industries and products. It is easy to recognize the forms of industrial growth periods. New building types, often large in scale, showing a number of technological innovations and built within a relatively short timespan, have a great impact on town plan and townscape – as seen in Bradford's Little Germany district. Such new industries drew workers, and new forms of housing were created for them and their families: in England these were courtyards and, later, terrace housing, but

FUNCTION: AN ECONOMIC AND TECHNOLOGICAL FORCE

43

elsewhere higher-rise flats or apartments were common. Housing was built as close to workplaces as possible: the intermingling of land uses in early industrial towns is very characteristic.

If settlements boom, can they also go bust? Yes – and we have already mentioned the impacts of rural depopulation. Any settlements that depend on one or two industries or trades for much of their employment can also suffer badly in periods of decline. There are many early ports from the mediaeval and even Roman periods which have declined sharply as their harbours have silted up. Hedon, now not much more than a village near Hull, is one example. Many villages in Wales and the north of England which depended on coal-mining have lost virtually their only source of employment as the mining industry has contracted. Symptoms of decline in key industries include large expanses of derelict land, vacant or under-used buildings, or their conversion to non-industrial uses such as retail or leisure. Large-scale unemployment caused by the collapse of key employers will result in physical symptoms of neglected dwellings as maintenance becomes too expensive, or vacancies as mortgage companies repossess. In extreme cases, as has been mentioned, a 'doughnut' effect can occur as property is simply abandoned. This can occur on both a large scale, particularly in 'company' towns, or a smaller scale in cities hosting a diversity of large employers.

Jacobs' important critique of planning focused on problem cities in the USA did discuss the symptoms of urban decline, but rather more in terms of non-physical aspects: community, economics, behaviour and organization [3]. Yet these factors easily, and often, interrelate to produce the startling physical aspects of decline: minimal maintenance, multiple occupancy, vacancy and neglect.

TRANSPORTATION: A TECHNOLOGICAL FORCE

Settlements cannot exist unless people have some mobility. And shifts in the availability of mobility provide, in all likelihood, the most powerful single process at work in transforming and evolving the human side of geography ... whose greatest measure is to be found in settlement, particularly in its most advanced form – in cities.[12]

Vance shows clearly that transportation, and particularly changes in its technology, have profound impacts both on urban form and on settlement networks. Changing transportation can cause the rise or fall of towns. Vance discusses the 'misplaced city' in the early rise of the US railroad system: the east coast harbour ports were well-developed cities which had lost the English trade after 1783 but were poorly located to exploit the American interior, unlike the river-mouth ports. Nevertheless it was these harbour cities which were behind the first phase of railroad development, in order to maintain their own prosperity [12]. So an understanding of the chronology and technology of transportation – turnpikes, canals, railways, motorways – can explain much about the settlement network.

Within the city, physical form has long been closely associated with transportation developments. Early technological advances, such as the canal, helped to concentrate urban functions on the transport nodes; hence the industrial concentrations around the canal networks of the 'Black Country' in the English Midlands. (It is alleged that Birmingham had more miles of canal than Venice!) Later advances began to allow a dispersal of urban structure, and cities typically developed out along radial axes followed by public transport routes. As this trend continued in inter-war England, fear of urban sprawl led to the 1935 Restriction of Ribbon Development Act and, later, to the restrictive Green Belts (see above).

Isard suggested that transport innovation and building development were closely linked – again, as with development cycles, in a cyclical manner. He noted that 'building represents more or less the culmination of the process of industrial, commercial, and population adaptation to the changing character of transport' [13]. For the USA, he identified the cycles of canal-building, peaking in the 1830s, railroads (three cycles: 1856, 1871 and 1887), the electric tramway (1906) and the first automobile cycle (1923, cut by the Great Depression). Each innovation involved great expense in the restructuring of existing settlements. The nature and scale of restructuring of Victorian cities to cope with the railway is well shown by Kellett, who discusses the commercial competition that led many towns to be served by several rail companies, each building lines and stations involving much property demolition [14].

Yet transport innovation arguably granted great freedom to urban populations. The US electric tramway allowed the first real suburban sprawl, permitting individuals to own separate dwellings with gardens, yet being able to commute to work in the city centre. The private car has continued this trend through the century, and cities have built motorway junctions and high-speed ring roads to cope with increasing vehicular traffic. But such construction takes time: Wolverhampton's ring road took well over 20 years and led to much blighting of property along the line of the proposed road before construction.

In the late 1980s, Birmingham felt that the inner ring road had become too much a 'concrete collar', restricting movement of people and expansion of the commercial core; and steps have been taken to break the collar, stopping vehicles to give priority to pedestrians. Elsewhere the menace of the car is being tamed with 'traffic calming' measures, and historic town centres are being freed from vehicles and pedestrianized. These may be relatively minor physical changes, but the visual and environmental impacts are great.

CULTURAL AND NATIONAL INFLUENCES

How can nationality or culture shape a settlement? Under this heading lie a complex set of historical factors which have shaped, and continue to exert immense impact upon, settlement form. The shape of the city in the western capitalist developed world, for example, is far different from that in the Islamic world (Example 2.3). Physical factors, including climate, underlie many of these cultural differences. Problems do occur when different cultural/historical influences interact as, for example, when colonial conquerors impose new settlements of new forms on a subject people. Such influences are well known in North America and Africa, but important examples also occur in Europe – the Norman conquest of England, the Moorish conquest of much of Spain, and the waves of Germanic and Slavic influence across Poland. The apartheid culture in South Africa also produced profound physical changes in settlements owing to the segregation of socio-economic and cultural groups on the basis of ethnicity.

Some of the most problematic examples of nationalist cultural influence and urban form have occurred in wartime. The razing of the Old Town of Warsaw in Poland by the Nazis was a deliberate attempt to subjugate not merely a rebellious town, but the sense of nationhood of a conquered people. The deliberate rebuilding, to a historic form predating the Germanic influence in this part of Poland, was an attempt to recreate national identity. This has been so successful that the new Old Town is now a major tourist destination and has been designated a World Heritage Site. Similarly, the rebuilding of Nuremburg (Figure 2.7) in a simplified mock-mediaeval style following British bombing was, according to Heinz Schmeissner, the Chief Planner, to preserve 'the concept of Nuremburg'. Although these rebuildings have shown success, the impacts of such cultural genocide are profound and long-lasting.

The conflicts of conservation in colonial countries following the departure of the colonial authority raise problems of whose heritage should be retained – the colonial or the indigenous?

The townscape dominants of San Juan [Puerto Rico] represented a proud Hispanic tradition which symbolized the Puerto Rican heritage and affirmed a sense of nationhood, especially after the

EXAMPLE 2.3

THE MOORISH TOWN

One of the key physical features of traditional Islamic towns is the dense network of narrow wandering alleyways. Public open spaces are small and irregular; through routes relatively rare; and there are numerous small culs-de-sac. In contrast to the western town, there has been throughout history a very different social structure. Sharply differentiated neighbourhoods developed, with segregation according to kinship, tribe or religious affiliation. The through routes divide such neighbourhoods, which themselves develop an inward-looking pattern of culs-de-sac and large internal open spaces – communal to the neighbourhood group, but not to the city as a whole. Such spaces could even be used for small-scale urban agriculture: the keeping of animals, growing citrus fruit, etc. The development of this Islamic urban pattern can most clearly be seen in those towns originating with a regularly planned Roman grid, such as Damascus or Mérida (Spain).

Figure 2.6 Figure ground plan of Ghardaia, Algeria (adapted from Benevolo, 1980, Figure 430). Streets and public spaces are black.

There were few rules for building design or settlement layout in the Islamic world. Streets had a minimum width of 7 cubits (about 3.5 m), allowing two laden camels to pass freely, according to a saying of the Prophet. Religious law and custom decreed the same dimension for the unobstructed minimum height of a street, to allow a person riding a camel to pass unobstructed. Standard public building types and styles – mosques, schools, markets – rapidly developed. But there was no urban government in the western sense. Key factors instead were the body of religious law and traditions, which addressed all aspects of public and private life; and the issues of ownership and privacy [15].

> You were not told what to do, what kind of city to design; you were only enjoined from doing things that threatened accepted social behaviour ... this concern asserted itself in the introversion of the house, the appearance toward the street being unimportant. At the same time, the traditional grouping of attached courtyard houses expressed a degree of interdependence ... [there was an] over-riding principle that older uses and established structures had priority over new uses and structures. Radical urban renewal was out of the question: urban repair, that is, piecemeal changes to the extant fabric, was the customary procedure.[16]

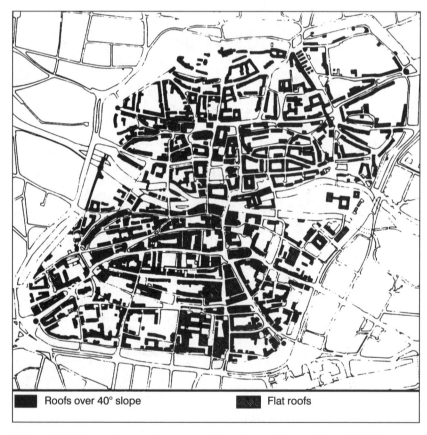

Roofs over 40° slope   Flat roofs

Figure 2.7 The roofscape of post-war Nuremburg (adapted from J.V. Soane, 1994).

EXAMPLE 2.4

CONSERVATION AND THE POST-WAR REBUILDING OF NUREMBURG [17]

Nuremburg suffered badly from Allied bombing during World War II. According to Georg Lill, the Conservator-General of Bavaria, writing in 1946:

> Nuremburg gave the spectacle of a huge pile of rubbish, a jumble of wires, pipes and installations and everywhere masonry carelessly cast about giving the appearance of disturbed mounds of earth instead of what we had hitherto known from the drawings of Dürer.

In considering its reconstruction, the most important priority was the recovery of the city's unique sense of identity. There were three possible courses of action to consider:

- an accurate stone-by-stone reconstruction of the mediaeval Old Town;
- a modern rebuilding in the form of rectangular streets and straight, high-density blocks of commercial buildings and residential apartments;
- a sensitive compromise using modern forms of construction but largely following the forms and proportions of the original structures, the mediaeval road pattern except for some essential widening, and retaining the pre-war location of the principal retail streets.

Although extensive areas of the Old Town had been completely destroyed, it was eventually decided to attempt the third option. One of the principal reasons for this decision, apart from the force of local tradition, was the existence of extensive remains of old structures throughout the destroyed areas. Another factor was the general ease by which the local building tradition could be adapted to suit the needs of modern society. The typical old Nuremburg house was a plain, unadorned structure of two to three storeys, sometimes with a projecting oriel window, but always with a steeply pitched roof and several attics. As long as the unique roofscape was reproduced, the proportions of the replacement houses could be properly adapted to accommodate as many floors and as many domestic alterations as were deemed appropriate. The lack of much traditional ornament was also an advantage to modern, efficiency-minded building techniques. External balconies would closely resemble another well-established, mediaeval architectural characteristic of the Old Town.

The chief deficiency in the reconstruction has been the considerable reduction of individual house plots (from 3040 m^2 in 1939 to 1360m^2 by 1970) in favour of the construction of larger houses on two or more amalgamated plots. Although this change has resulted in a certain reduction of absolute visual authenticity (the mediaeval plan has been considerably simplified), it does not detract from the general success of the overall appearance of the reconstruction. Although the façades of individual buildings risk being plain and monotonous, in general terms the four major aims of the rebuilding of the Old Town – the preservation of major monuments, their integration into a sympathetic urban environment, the preservation of the essential layout and social structure of the mediaeval centre, and the effective isolation of the latter from the rest of the city – have been achieved and the city is, once again, successful as a tourist destination.

Nuremburg is therefore a good example of reconstruction along old patterns following a catastrophe. Although slightly simplified, the key physical characteristics – the skyline of steeply pitched roofs seen from the castle, for example (Figure 2.7) – have retained or recreated the sense of identity.

United States asserted sovereignty at the end of the nineteenth century. It was clearly in the national interest to preserve the symbols of Puerto Rico's heritage ... The old city core evinced a strong sense of community continuity over several centuries that long predated American influence in the region.

The patrimony of the Old Stone Town of Zanzibar, however, contained no symbols of the indigenous Swahili culture but represented the culture of Arabs, Indians and, to a lesser extent, Europeans. The landmark buildings in the Stone Town loudly proclaimed the wealth, power and domination of the colonial era ... [Conservation was] to halt the collapse of housing, revitalize the local economy and so preserve the urban patrimony of the Stone Town as a testament to the diverse origins of the population of Zanzibar, notwithstanding the recent painful memories which many of the structures evoked.[18]

FASHION: A CULTURAL FORCE

Changing fashion is a significant but under-regarded factor in shaping settlements. As with many other aspects of design, from cars to clothes, changes in urban fashion may be rapid and wholesale, although some places will remain relatively little affected by any one change.

One clear fashion now dominating North America and England, but much less marked in Scotland and much of Europe, is the trend towards suburban living. The individual house, preferably detached or semi-detached, and its garden are dominant in suburbia; and people clearly prefer this type of dwelling. Tenement living is much more accepted in urban Scotland, while in continental Europe high-density dwelling in the urban core is still usual, perhaps a relic of the more recent warfare between city-states and the need to live within urban fortifications. In these cases, many people also have 'leisure gardens' or summerhouses, often some distance from the city itself. Yet the trend towards the individual house is becoming apparent; for example, in the large numbers of incomplete self-built houses appearing in rural areas in Poland and new bungalows in Malta.

Architectural style is another important example of fashion. Here, fashion is dictated by a mixture of the designer, client, finance provider and market forces. The significance in terms of the urban fabric can be seen in the progression in many UK towns from various styles based on historical ideas (e.g. neo-Georgian) to the post-modern over the last six decades.

In commercial developments in UK urban centres during the twentieth century, the setters of architectural fashion have been increasingly non-local, especially large London-based firms that are driving out the local smaller developers and architects who shaped urban areas in the interwar and earlier periods. To a considerable extent, they introduce novel architectural styles, especially Art Deco in the interwar period and Modern in the post-war period. These styles are then adopted and further diffused by local firms only after some delay. An exception to this principle is the occasional owner–occupier developing for themselves, who may seek to create a new impression with an eye-catching novel style. During the last two decades, the ubiquitous Modern style has become displaced, particularly in commercial architecture, by 'post-Modern', characterized by the use of historicist styles, local idiom and materials, and which in general relates far better to existing British urban contexts than did the Modern style. It is evident that post-Modern developments in North America are just as large and visually obtrusive as their predecessors; in Britain, as with Art Deco, the style has generally been adopted in a rather more restrained manner.

How do these trends appear in suburbs? Recent studies of UK middle-class suburbia, particularly in the interwar period, have shown how the originally élite tastes shown in aristocratic country house designs filter down the social hierarchy and are catered for by speculative builders – the 'stockbroker Tudor' caricatured by the poet and architectural journalist John Betjeman is typical. In the 1970s, a form of Georgian was popular – almost any standard housebuilder's 'box' had small-paned windows fitted and was marketed as neo-Georgian. In the USA, a 'Queen Anne' style is currently very popular in high-class exclusive suburbs.

Developing a knowledge of how architectural styles, and also other factors such as residential estate layouts, have varied over time enables us to visit a new settlement and, almost immediately, begin to develop an idea of what parts were developed at what period and, sometimes, what types of agent were involved.

It should be realized that fashions in physical form are closely related to fashions in professional attitudes, themselves closely linked to professional education. For example, large-scale cycles have been observed in the amount and type of town planning activity in Britain. There have been upswings in 1890–1914 and 1945–1970, with downswings between 1918 and 1939, and 1970 to the present. Such cycles are closely related to, and difficult to distinguish from, cycles in the free-market uncontrolled economy. The upswing after World War II brought about the New

51

Towns and, allied with the prevailing Modernist fashion in architecture, reshaped many town centres. In the majority of built environment professions at present, the dominant fashion is for conservation, rehabilitation or refurbishment, rather than for comprehensive clearance and

EXAMPLE 2.5

CHANGING ARCHITECTURAL STYLES IN WORCESTER

In a study of development in the core of the historic cathedral city of Worcester, it has been possible to classify most new buildings into three broad groups: Modern, neo-Georgian and post-Modern.

Forty buildings were categorized as Modern, their dominant features being façades based on horizontally sliding windows or glass curtain walls; a certain recurrence of regular forms but with irregular volumes; the use of glass, artificial stone or exposed concrete as main building materials; and a lack of harmony between the design of the new building and surrounding buildings. A few of the 17 buildings categorized as neo-Georgian are reasonably faithful reproductions of the main characteristics of genuine Georgian façades. True Georgian architecture is very common, if not dominant, in the commercial core of Worcester, and these buildings could be classified as Georgian revivals. In the majority of cases, there is clearly more emphasis on integrating the new building into its urban context than on stylistic accuracy. In these cases, the façades continue some of the main characteristics of the local eighteenth century architecture, principally the use of brick, regular spacing and shapes of windows, and moderate building volumes. The new buildings do not merely reproduce Georgian originals, particularly not in the case of typical Georgian architectural details. These buildings are neo-Georgian, not revival Georgian: this fine distinction is of some interest.

The concept of post-Modern architecture is used in its widest sense for 10 buildings which demonstrate the eclectic use of classical or vernacular architectural elements, together with modern ones, without reaching the point of being a simple reproduction of any historical style.

The more significant alterations to listed buildings have also been included.

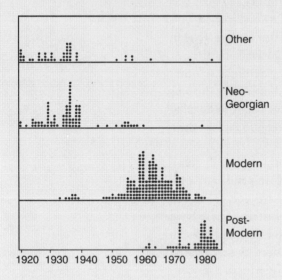

Figure 2.8 Changing architectural styles in Worcester (Vilagraa, J. and Larkham, P.J.).

Figure 2.8 shows the number of developments carried out, analysed by architectural style and year of planning permission. The first point of note in this figure is the significance of development in the 1960s, and its coincidence with the clear predominance of developments which used Modern architectural styles; these were important locally between the late 1950s and the early 1970s. Many of these were developed by retail chains operating nationwide. Coexisting with these Modern developments, those new buildings adapted to the predominantly Georgian surroundings stand out, although quantitatively they are of lesser importance. Revival styles, particularly neo-Georgian, were of some importance in the United Kingdom in the inter-war period but did not appear in significant numbers.

redevelopment. If these fashions are cyclic, then it is likely that this conservatism, or conservation-consciousness, will not be dominant for ever!

WORKPIECE 2.3

PLOTTING CHANGING FASHIONS

You will need a large-scale map of a settlement (or part of it). Town centres are preferred.

Cover the map with a sheet of tracing paper and, using different colours and/or shading, plot the various architectural styles present (these, of course, will be different in different countries). Wherever possible, put dates to buildings. Be aware of 'revival' architectural styles (see Example 2.5).

Are there any parts of the settlement where one style, and date, predominates? This would be unusual, and suggests a town formed largely at this one period, followed by economic stagnation (little pressure for change).

Are there examples where revival styles are located next to 'genuine' examples? What dates? If recent, this might indicate a concern for conservation – the character and appearance of areas and buildings.

PEOPLE AND PERSONALITIES: A CULTURAL FORCE

Settlements are lived in and shaped by people, and the impacts of both individual personalities and group dynamics can be powerful. An old townscape is the product of the aspirations and actions of many generations. Some of their actions persist into today's townscape; of others, little or no significant trace remains. What factors, other than those already discussed, drive people?

One significant factor is ambition. Clearly, some people are motivated by the acquisition and retention of power; for example, through the local or national political systems or through advancement in a profession. Power can be exercised in a number of ways which might have consequences for the built environment. This is particularly true when power has been concentrated into few hands. The form of post-war Birmingham was significantly affected by the influence of Herbert Manzoni, chief officer for the city's Public Works Committee and a civil engineer. His training and vision led to an inner, middle and outer ring-road system, demolitions for which removed thousands of slum houses; recently the inner ring road has imposed unforeseen constraints on the growth of the city core. Yet such concentrations of power are often only apparent when corruption is proved, as was the case in Newcastle-upon-Tyne in the 1960s with T. Dan Smith and Poulson: again their developments, such as the giant Eldon Square shopping centre, radically changed the city centre.

EXAMPLE 2.6

HAUSSMANN'S PARIS: A CONCENTRATION OF POWER

The largely mediaeval and baroque form of Paris was greatly altered in the period 1851–1870 under the rule of the Emperor Napoleon III. He allowed one key individual, Baron Haussman, to take charge of the large-scale alteration of the city to provide an up-to-date national capital. Haussman was not only skilled in designing the new street layouts and controlling the shape and style of the new buildings that lined them: he was also adept at using existing legislation – particularly a law allowing compulsory purchase of property and another concerning public health – and also at raising and managing the enormous sums of money necessary to fund the rebuilding.

The main changes included the following:

- New streets laid out within the old city. Haussman added 95 km of new streets to the existing 384 km network; he also removed 50 km of old street lines to make way for the new. This new street plan was integrated with the baroque boulevards, which had been built on the site of the old defences.
- New streets in the surrounding area. An additional 70 km of streets were laid out outside the old city.
- New infrastructure, including water supply, sewers and gas lighting.
- New public facilities, including secondary schools, hospitals, prisons, etc.; and public open spaces in the Bois de Boulogne (west of the old city) and Bois de Vincennes (to the east).

These works cost the city some 2.5 billion francs. The city's population doubled during the period, to about 2 million.

Haussman endeavoured to improve the quality of the new environment ... he tried to impose a degree of geometric regularity and chose some form of monumental structure, either ancient or modern, to provide the focal point of each new street. He enforced the architectural uniformity of the façades overlooking the most important squares ... [But] the different areas lost their individuality and blended into each other, the façades of the buildings became merely a constantly unfolding backcloth, and the street furniture – lamps, kiosks, benches, trees – began to assume a much greater importance.[19]

Another (contemporary) view of the scale of change was also critical:

...the old Paris, struck in the heart by Baron Haussmann, never rose again. Of the old picturesque city ... hardly anything remains save a few hotels dishonoured by advertising placards, a few blocks of stinking houses, a few notorious alleys.[20]

Nevertheless, much of the present character and appearance of Paris – its street network and many buildings – owe a great deal to Haussman. The scale and speed of change was made possible by a government sufficiently powerful and autocratic to conceive and to carry out work on such a large scale and by finding in Haussmann someone well able to design and manage the project.

MEASURING THE IMPORTANCE OF PERSONALITIES

Using a street plan or A–Z plan, examine any built-up area and note down any personal names. Who is commemorated? Can you suggest the date of the streets so named? Do you recognize any of the names from the area's history?

Local politicians, MPs, wealthy families and so on are often remembered in street or even building names.

In the Victorian streets of terraced houses, we can often find names of the builder's family: in Birmingham there are Emily Street and Louise Street dating from about 1895.

Remember that there are also other fashionable names – the Nelson Mandela place names of the 1980s, for example.

Politics is an important factor in settlement change. Most developed countries have a permit system to regulate development, and most permit decisions are made by local politicians (or the national government if the issue is important enough). Large-scale impacts of political decisions can be seen in the UK's post-war system of new and expanded towns, set up to cope with increasing overcrowding in the major conurbations, and resulting in completely new settlement designs usually characterized by residential 'neighbourhood units' and the segregation of pedestrians and vehicles. Small-scale political influence is seen in site-by-site planning decisions and, for example, in the controversy surrounding the involvement of local and national government, businesses and local pressure groups in replanning Georgian Dublin over the past two decades. Here, politics and personality were influential in directing redevelopment proposals, new road schemes and funding, initially at the expense of what was seen as the 'colonial heritage' of Georgian Dublin [21].

On a smaller scale – the more familiar and everyday scale of changes to individual dwellings and their plots – the important personal factor is the stage of the individual or family life cycle. As we grow older, and children become independent and set up their own households, our needs from any property change. To cope with children, a house may be extended. Later, this may be turned into a self-contained 'granny-flat'. Later still, a large garden may be too difficult to maintain, so it is either subdivided and part sold, or the owners move to a smaller property. It is this family life cycle which determines the great majority of individually small-scale changes to residential buildings.

At what speed do all these changes occur? The majority of changes are relatively small and occur relatively infrequently. Although a high-street shop may change its store layout and signs every five years or so, a family

THE RATE OF CHANGE

may alter its house much less frequently. The majority of changes are small scale, so the average town changes only gradually. However, there are some occasions when catastrophic change occurs. This may be due to war, accident or natural catastrophe. The Great Fire of London in 1666 devastated the city centre and allowed many idealistic reconstruction proposals to be put forward: the best-known is that of Sir Christopher Wren although, in fact, the city was rebuilt to a plan very similar to its pre-fire state. Haussmann's rapid changes in Paris were also seen as a catastrophe (Example 2.6).

Rapid changes can also occur when there are particularly powerful agents directing the change. These may be individuals, companies or local authorities. Again, Haussmann's Paris is a relevant example.

The dominant UK post-war fashion of comprehensive clearance and redevelopment, strongly supported by many UK local authorities, led to immense and rapid change [22].

CONCLUSION

The main message of this chapter is that understanding settlement origins and growth is a very complex topic, but one necessary to further understanding of factors affecting the built environment.

Clearly, there is considerable diversity of processes, actors and forms involved in shaping settlement form. A good understanding of history is very helpful in understanding the particular factors active in any one place; and knowledge of comparative studies – similar processes operating in other places – allows us to make suggestions even when the exact details remain unknown.

We can begin the process of studying urban form by looking at representations of forms on maps – preferably a series of historical maps for the same area – and at current conditions on the ground. From a basis of historical knowledge and comparative studies we can begin to deduce likely processes shaping these forms and the variety and nature of the actors involved. Ideally, we can then begin to confirm these deductions from other sources, such as historical documentation. Since such documentation does not always exist, particularly for older periods, we must sometimes keep to our informed guesses!

The persistence of settlement form – whether street networks, plot patterns or building types – can lead to both constraints and opportunities. A recognition and understanding of form greatly assists in resolving constraints and taking advantage of opportunities. Currently, conservation is a high-profile issue involving persistence of forms. The deliberate retention of buildings and areas can lead to significant inward investment and regeneration centred around a conserved structure, as with London's Covent Garden or Boston's Faneuil Hall. A preserved building can also stand in the way of necessary redevelopment – for example, if a new or wider road is necessary. Furthermore, retention does not necessarily lead to renovation and re-use: even preserved buildings can be neglected and demolished.

'Cities are an immense laboratory of trial and error, failure and success, in city building and design'[3]. We need to be able to identify and understand the reasons for success and failure; to learn lessons from the past.

The physical form of settlements of all types and all dates is the key product of the development process. Its production has involved, in some degree, all of the built environment professions, but it is also consumed, or used, by all of us. An understanding of how these forms came about is an invaluable part of the education of all involved in the production of new settlements or parts of settlements.

This understanding demands some knowledge of a wide variety of factors which operate at different scales and intensities in every settlement. These factors will also change over time, but some are cyclic and so they recur again and again.

This chapter has:

● shown why an understanding of settlement form is important;
● developed skills in analysing settlement form;
● considered some of the key factors involved in shaping settlement form.

The issues covered in this chapter are:

● how to recognize key features of the physical form of settlements;
● how to recognize physical evidence of periods of growth and decline;
● how to develop an understanding of the range of processes and actors shaping settlement form;
● how lessons can be learned from past urban forms to help shape future developments.

SUMMARY

CHECKLIST

REFERENCES

1. Brolin, B.C. (1980) *Architecture in Context*, Van Nostrand Reinhold, New York.
2. Carter, H. (1990) *Urban and Rural Settlements*, Longman, London.
3. Jacobs, J. (1961) *The Death and Life of Great American Cities*, Random House, New York.

4. Conzen, M.R.G. (1962) The plan analysis of an English city centre, in *Proceedings of the IGU Symposium on urban geography, Lund, 1960* (ed. K. Norborg), Lund Studies in Geography series B 24, 383–414.

5. Slater, T.R. (1990) English medieval new towns with composite plans: evidence from the Midlands, in *The Built Form of Western Cities* (ed. T.R. Slater), Leicester University Press, Leicester.

6. Vilagrasa, J. and Larkham, P.J. (1995) Post-war redevelopment and conservation in Britain: ideal and reality in the historic core of Worcester. *Planning Perspectives*, **10**(2), 149–172.

7. Gottmann, J. (1978) *Forces Shaping Cities*, Department of Geography, University of Newcastle upon Tyne.

8. Whitehand, J.W.R. (1987) *The Changing Face of Cities*, Blackwell, Oxford, Chapter 2.

9. Lewis, J.P. (1965) *Building Cycles and Britain's Growth*, Macmillan, London, p. 22.

10. Whitehand, J.W.R. (1988) Urban fringe belts: development of an idea, *Planning Perspectives*, **3**(1), 47–58.

11. Castells, M. (1977) *The Urban Question: a Marxist Approach*, Edward Arnold, London.

12. Vance, J.E., Jr (1986) *Capturing the Horizon: the Historical Geography of Transportation*, Harper & Row, New York, pp. 2–3.

13. Isard, W. (1942) A neglected cycle: the transport-building cycle, *The Review of Economic Statistics* **24**(4), 149–158.

14. Kellett, J.R. (1979) *Railways and Victorian Cities*, Routledge and Kegan Paul, London.

15. Hakim, B.S. (1986) *Arab-Islamic Cities: Building and Planning Principles*, Kogan Page, London.

16. Kostof, S. (1993) *The City Shaped*, Thames & Hudson, London, p. 63; also citing Hakim (1986).

17. This Example has been adapted from Soane, J.V. (1994) The renaissance of cultural vernacularism in Germany, in *Building a New Heritage* (eds G.J. Ashworth and P.J. Larkham), Routledge, London.

18. McQuillan, A. (1990) Preservation planning in post-colonial cities, in *The Built Form of Western Cities*, (ed. T.R. Slater), Leicester University Press, Leicester, p. 405.

19. Benevolo, L. (1980) *The History of the City*, Scolar Press, London, p. 798.

20. Eschollier, R., quoted in Olsen, D.J. (1986) *The City as a Work of Art*, Yale University Press, New Haven, p. 44.

21. McDonald, F. (1989) *Saving the City: How to Halt the Destruction of Dublin*, Tomar, Dublin.

22. Larkham, P.J. (1995) Constraints of urban history and urban form upon redevelopment, *Geography* **80**(2), 111–124.

Aston, M. and Bond, J. (1987) *The Landscape of Towns*, Alan Sutton, Gloucester (paperback reprint).

Benevolo, L. (1993) *The European City*, Blackwell, Oxford.

Burke, G, (1971) *Towns in the Making*, Edward Arnold, London.

Drakakis-Smith, D. (1987) *The Third World City*, Methuen, London.

Ford, L.R. (1993) *Cities and Buildings*, Johns Hopkins University Press, Baltimore.

Relph, E. (1987) *The Modern Urban Landscape*, Croom Helm, London.

Stenhouse, D. (1980) *Understanding Towns*, Wayland, London.

Whitehand, J.W.R. (1991) *The Making of the Urban Landscape*, Blackwell, Oxford.

MODERN URBAN PLACES

TOM MUIR

THEME

The development of our great historic towns and cities is often regarded as a happy accident but some places seem more accidental and much less happy. Rapid technological change since the industrial revolution combined with the emergence of new approaches to planning and architecture have often resulted in urban landscapes and environments of variable and sometimes very poor quality. Changing transportation systems and development patterns in growth and decline have created diverse and often 'disorganized' townscapes. This chapter examines the characteristics, both good and bad, of modern cities on an international scale with particular reference to the European experience. It seeks to explain how their changing development patterns have influenced the nature of space in the modern city. With these themes in mind, the chapter presents a review of the different qualities of space found in cities and attempts to relate these spatial qualities to the main activities of the city's commercial, residential, industrial and retail functions by exploring the relationship between these functions and the urban spaces they generate, and by indicating some aspects which are negative and some which are positive. Examples are used to suggest how places work and how they have evolved through time, and to explore the suggestion that not all public squares are exclusively dependent upon the quality of either the architecture or the space itself.

OBJECTIVES

The objectives of this chapter are:

● to develop an understanding of the relationship between urban activities and their attendant places;

- to provide some explanation for the distribution of such space throughout the city region;

- to identify some internationally recognized squares/places and establish how long it took for them to evolve into their present form.

> The street, which is often thought of as a channel for motion, was almost as often a fixed place – a site where things happened. People gathered in the street to talk or to shop, they ate at tables set outside its cafés: children played in it, old people sat on benches. Even more clearly the square was a location where one was supposed to be at rest. Approached through the city's byeways, the square formed a natural destination or stopping place. Let the street be widened to a boulevard, however, let the boulevards evade the square's seclusion – now a whole new kind of space comes into being. This space is no longer passive. It exists only when we keep going; it is sensed only in motion.[1]

This view expresses clearly the dilemma of space in the modern city. The increased pace of movement, both through and within, has led to the development of just this situation. Roads, hitherto community foci, are now community divides. Squares and places which were previously collectors of people are now traffic islands. Why has this happened? Is it an inevitable process of change redefining the city for its new, evolving purposes?

In this chapter urban space will be examined through the study of four of its seminal functions: residential, retail, commercial and industrial. These central functions are supported by a vast network of services such as roads, railways, education, recreation and many others, all of which combine in infinite complexity to create a working city. Space between the buildings, whether planned or residual, is the key catalyst to the quality and effectiveness of the total results.

> The building of cities is one of man's greatest achievements. The form of his city always has been and always will be a pitiless indicator of the state of his civilization. This form is determined by the multiplicity of decisions made by the people who live in it. In certain circumstances these decisions have interacted to produce a force of such clarity and form that a noble city has been born.[2]

CHARACTERISTICS OF MODERN CITY DEVELOPMENT

As the modern city evolves, each site is developed according to the needs of the user and in relationship to the infrastructure which is required to support its use. This means that the changing nature of society is reflected in its physical development and never more so than in the western world in the last 15 years, during which a powerful free market ethic has prevailed. Let us now examine how some of the most common types of development have contributed to the spatial characteristics of the modern western city.

COMMERCIAL

Undoubtedly the fastest growing building type, the office block, has become the archetypal image of the twentieth century city. Made possible by the new technologies of glass, concrete, steel and the fast vertical transit system, the elevator, its ubiquitous form symbolizes the new urbanity. The office block's rise to prominence is coincidental with the drift away from industry as the prime activity within the city. Commercial activities command the highest rents, and thereby result in the highest land values[3] thus giving it prime right of choice in site selection. This also necessitates, however, a high density use of the site – hence point blocks sometimes 50–80 storeys high based upon an incredibly small plan form. An example of this is Centre Point in Camden Town, London.

This small plan form, generated partly by land costs and partly by the desire to minimize deep plan, while retaining lettable space, resulted in the fragmentation of space found at ground floor level. Many of the original tall office blocks rose straight off the pavement and the character of cities such as New York and Chicago is a direct result of the chasm-like streets between high blocks imposed on a powerful grid-iron plan. Little contribution to public space was made by these buildings, but there are three classic examples of attempts to achieve this.

Figure 3.1 The Lever Building, New York (Skidmore, Owings and Merrill, 1952) – tall block set back behind low podium.

- The Lever Building, New York, was designed by Skidmore Owings and Merrill in 1952 (Figure 3.1). They created a low two-storey block to the edge of the site to create a covered public 'square' underneath. The main 20-storey office block was set back at the rear of the site[1].
- The Seagram Building, designed by Mies van der Rohe and Philip Johnson in 1958 placed a tall 37-storey office block well back on the site in order to create an open public plaza in front (Figure 3.2).
- The Rockefeller Centre, by Wallace and other architects, created a complex of 13 buildings designed around a public square (Figure 3.3). The buildings were designed to complement each other and enhance the square[1].

Figure 3.2 The Seagram Building, New York (Mies van der Rohe, Philip Johnson, 1958) – public space created by setting 37-storey office block to rear of the site.

Figure 3.3 The Rockefeller Centre, New York – public square created by grouping development of tall office blocks around open space.

In each of these cases, a clear decision was taken to leave some of the site unbuilt in order to attract pedestrian activity and introduce the vitality at street level so often missing where the building rose directly from the pavement.

In London, an example of such an approach is the Economist building by Peter and Alison Smithson (1964). This plan created a pedestrian route through the site between two tall office blocks which was intended to encourage a lively flow of people at ground floor level both inside and between the buildings.

In each case mentioned so far there has been an attempt to regain some space for the general public from a building type that is substantially for private use and which has come to dominate so much of our downtown areas in the latter part of this century. Commercial development has also been moving out from the central business district and establishing peripheral suburban business zones. These are the result of high, dense office developments requiring more attractive and spacious access, as well as a greater facility for parking. This pattern is largely a response to the rapid increase in the use of private cars, particularly in the relatively well-paid sector of the community from which office workers are mostly drawn.

The geometrical pattern of site use created by such developments is predominantly influenced by the needs of the motor car. The pedestrian usually has to negotiate a combination of car parks and access roads, before arriving at the main, and normally the only, public entrance. The landscape treatment of the entrance is often of high quality, but developers tend to serve their own buildings exclusively and the public domain is frequently ignored. Sidewalks or pavements are disjointed and often the landscaping of the spaces seems to be designed more to impress visually than to serve as an effective public place.

Perhaps this suggestion of private space is the key to one of the major problems associated with office developments. Their activities are by their very nature private and therefore resources are primarily directed towards satisfying these private needs. Occupation of the building is probably only during office hours and this is normally for seven to eight hours per day and for only five days per week. The remainder of the time the building neither makes demands from, nor contributes to, the vitality of the surrounding area. In districts such as the City in London or in Wall Street in New York, this characteristic is all too obvious with largely deserted streets and empty spaces after office hours and during weekends.

Another manifestation of this condition is the phenomenon called the **doughnut effect** so often to be found in North American cities, whereby the downtown area is predominantly populated during the day with office workers but is deserted at night. There are, however, other reasons complementary to the 'nine-to-five' office block which have contributed to this development. The characteristics of modern urban spaces are conditioned by the nature of our urban society and the changing life styles of city dwellers. Shopping and housing, two other activities central to urban life, are undergoing fundamental changes both in their location and in their character[4].

The continental tradition of living in the city centre is one most strongly associated with those cities which have pre-industrial origins. Large European cities such as Milan, Frankfurt, Barcelona, Amsterdam and many more have since mediaeval times had a substantial proportion of their population living in the centre, and so the entire city is constantly populated. In the industrial city, and now in the post-industrial city, those mediaeval origins can be seen in the degree to which their present populations still enjoy living in their centres. The loss of this residential role has provided the greatest threat to the vitality and excitement of our post-industrial cities. A resident population demands a range of services such as schools, shops, trades, employment, etc., which ensures a lively, round-the-clock range of functions. Such a city never really dies, nor does it become zoned into a series of more functional districts each of which has a limited period of activity. Housing provides a widely diverse range of uses for the spaces in the city. Streets become meeting places and small shops, cafés and bars open to provide locations for such meetings [5]. Squares become focal points for activities and market-places during the day; at night they become social centres, often taking on a completely new but vital life. Even car parking, which overwhelms so many downtown streets and squares, is at least indicative of a desire to come into the city for entertainment and social life.

The great British industrial cities including Manchester, Sheffield, Leeds and Birmingham, in common with most cities in North America, have never had an inner-city residential tradition. In Britain, the industrial revolution in the eighteenth and nineteenth centuries polarized residential districts. The wealthy industrialists usually chose to live in the west end of the cities (up-wind of their factories); the factory workers and urban poor lived in a tight ring around their places of employment, in slum conditions. Thus, apart from certain more salubrious dis-

RESIDENTIAL

tricts in London (such as the urban estates of the aristocracy), inner-city living became associated with poverty and slums. There was little or no purpose-made public space in such residential ghettos, although they benefited to some extent from the monumental civic centres and public parks which grew up in the late nineteenth century, generated through the growth of powerful local councils.

In the absence of such space, the street outside the long rows of industrial terraced houses took on a new role. The houses were very small, with minimal yards at the back and front doors opening directly from the front room on to the back of the pavement. The street therefore became almost a common front garden or yard for all the residents. It was the place where the community met and exchanged talk and created a social network. This type of place may enable its users to develop a sense of territoriality and defensible space in which each occupant plays a role in overseeing the street.

This conversion from 'channel' to 'place' is also evident in some of the major arterial routes in cities. For example, Birmingham is a city of radial roads taking the traffic right into the heart of the city. Two such roads demonstrate how different their characters can be depending upon the surrounding community. The Hagley Road (Figure 3.5) brings traffic in from rural commuting villages and has been widened to facilitate rapid transit. The housing which had existed on either side has withdrawn from it, forming two quite different communities and leaving the road to be flanked by hotels, restaurants and office blocks, often in the original converted houses.

In contrast, Soho Road (Figure 3.6) connects an adjacent town centre in the West Midlands conurbation with Birmingham and passes through a lively multicultural community. This road has become the marketplace for the surrounding communities and is the common element that unites both communities on either side. Through-traffic has bowed to the inevitable and travels at a pace commensurate with the local uses. The battle for this road is being won by the community: traffic that wishes to move faster simply chooses another radial route.

Figure 3.4 Typical street of late nineteenth century workers' houses opening directly on to the pavement.

Figure 3.5 Main commuter road into city centre – priority to vehicles.

Figure 3.6 Community-orientated radial road into city centre.

WORKPIECE 3.1

STUDYING TRANSPORT CORRIDORS

Major arterial roads in a town or city can either be incorporated into the communities through which they pass, or be divisors, splitting the local population into two separate communities.

Choose a major city road you know that goes from the edge to the centre of the settlement. Analyse its character through the length of its route, identifying in particular how it changes as it passes through different sorts of area.

Look at it both through the eyes of the road user and through the eyes of the different communities. Identify in particular the character of the places created by the different types of buildings and patterns of development.

Perhaps the greatest change in residential planning came with the housing estates of the post-war period (Figure 3.7). These developments relocated inner-city residents into predominantly public housing both within the city where their previous terrace housing was demolished or in large peripheral housing estates. These were laid out in a geometrical pattern with streets no longer being the generators of the urban form and with vehicles and pedestrians separated. The predominant house type was the flat, or apartment, and these were grouped in blocks ranging from five-storey terraces to blocks of 20–30 storeys.

The sites were landscaped with access for vehicles and car parking and extensive areas of public open space. These again were generally ill-defined and lacked a sense of ownership. One infamous site was defined as being for 'public amenity, no ball games to be played'. Residents, moved from their traditional nineteenth century streets, found no facility for the community focus role of the street nor any effective replacement. The public space around the blocks was substantially neutral, with none of the residents (including those living in ground floor flats) offering supervisory or even observation functions. Small shopping precincts were built, but the lack of any identity or ownership of the spaces in the development resulted in their rapid deterioration along with the covered car parks and, eventually, many of the flats themselves.

Figure 3.7 Post-war (1960s) council housing estate.

Why did this apparently generous planning not work? Clearly, no real relationship developed between the flats' residents and the landscaped site surrounding their blocks. People discovered that their private territory ended at their front door and that there was no possible way for them to 'police' the public spaces which came right up to their blocks. These became totally public and, being so close, were considered to pose a threat to the residents' security. Perhaps these reasons, more than any, led to the movement for area rehabilitation rather than redevelopment which has gathered momentum since the early 1970s.

WORKPIECE 3.2

CONTRASTING HOUSING DEVELOPMENT PATTERN

Twentieth century housing estates generated a new pattern of geometric planning and layout in the UK and North America – both of buildings and the spaces between.

Compare such housing layouts with those of previous periods, particularly the nineteenth century.

Consider the amount and quality of the public spaces created, and the nature of private spaces and their relationship to the surrounding houses.

SHOPPING

While the mediaeval town was once described as being one single marketplace, the commercial reality of shopping can still be seen to occupy a pivotal role in the modern city. There are, however, several fundamental developments in shopping behaviour which in some ways might be seen as the greatest threat yet to our traditional city centres.

Through industrialism to post-industrialism, retail trade has developed as one of the most vital aspects of the central business districts in towns and cities. As first industrial and then commercial activities sought more space and moved out to the edges of town, shopping and entertainment replaced them as the prime role. The traditional form was the street – with names like Regent Street, Fifth Avenue, Rue Rivoli and the Kusfurstendam representing many more such shopping streets of international reputation. Shops along the sides, wide pedestrian pavements and traffic in the middle – a lively, vital place in which social as well as shopping activities took place. This is now under severe threat, not so much in those European towns with strong inner-city residential traditions, but in Britain and North America where there is no such strong tradition of living in town and city centres.

The traditional relationship of shops and public space is being replaced by incorporating them into 'malls' which are private developments. These malls act rather like department stores except that each

Figure 3.8 The Galleria – large shopping mall outside Dallas, Texas.

shop inside is a private unit and the mall-owner manages the overall space, which in many cases, given that sometimes 100–200 shops are enclosed, is substantial. Why do these malls pose a threat to downtown shopping?

- They open and close like one store; therefore the entire development has controlled entrances and, apart from a few situations where they must provide access to a public amenity such as a railway or bus station, they open only during daytime shopping hours.
- They frequently employ a private security service. While this may be looked upon as a benefit for personal security, it is possible to see the growth of private policing on this scale as a process which privatizes the public realm (Figure 3.8). Will this trend lead to a polarization between wealthy 'safe' places and poorer 'unsafe' places?
- Sometimes as much as 20% of available retailing in a town centre may now be in such malls, reducing the opportunities for evening and Sunday window-shopping. Is the publicly accessible town centre diminishing?
- Finally, and most significantly, malls have decided that they no long need to be in the town centre. Indeed, by using the cheaper land available in the surrounding countryside, they can create a whole new concept of shopping and recreation with free car parking and bus ferry service from adjacent towns.

What is the quality of place that is being created in a modern, out-of-town mall and that is proving so attractive to shoppers in comparison with the traditional town centre? There are four advantages:

- controlled climate (no rain/snow/cold);
- traffic-free shopping (ideal for families and the elderly);
- high in entertainment (lots of events and activities);
- free, accessible car parking.

Then there are some disadvantages:

- There are few real surprises, with all experiences of an even consistency – no peaks, no troughs. It is a shopping equivalent of fast-food.
- The environment diminishes the intellectual stimulus. Everything is packaged to stimulate the shopper to spend money.

● There is no sense of 'place'. Shopping malls are purely functional and more like a stage-set than a real town. They have no history or traditions.

Perhaps the malls' biggest problem is that the advantages are mainly in the area of efficiency and function and they fit well into a certain popular lifestyle. This can have consequences elsewhere. For example, the success of Merry Hill in Dudley, West Midlands, England, is having an extremely damaging effect on the economic viability of the local towns of Wolverhampton, Dudley, Walsall and Kidderminster and is even threatening major shopping centres such as Birmingham (Figures 3.9 and 3.10).

In America, malls are a fully established reality and in most large towns or cities out-of-town shopping malls are the only place to shop, as they contain all the major stores. They have had a disastrous effect on downtown areas and are substantially to blame for the evening and weekend desertion of the city. They do, however, prove extremely popular and the newest of these giant centres are starting to broaden the range of services they offer. By grouping with other large out-of-town facilities like DIY centres, furniture retailers and garden centres, a new type of decentralized town centre is the part of the phenomenon which most threatens our traditional concepts of urbanity and, most relevant to our subject, urban space.

In this new urban form, urban design and architecture have been replaced by interior design and stage-setting. The buildings themselves make little pretence at an architectural statement, nor does their position in relation to each other create successful urban space.

The only point at which the buildings make any attempt to communicate externally is at their entrance. It is upon passing through this point that this new form of urban 'place' is encountered. Not quite a building, not quite a street, it is this ambiguity linked with the brightly lit, colourful interior all at an ambient temperature, whatever it is like outside, that makes it popular. It is, in fact, a microcosm of the old mediaeval market town brought up to date with modern technology. There is, however, a fundamental difference. Whereas the entire life style of the mediaeval town dweller revolved around marketing, the shopping mall represents one element of modern life style: consumerism. Shopping has been extracted from its traditional place, where it was integrated into a multifunctional town centre, and has instead been grouped as a concentrated, single-function development devoted exclusively to retailing and those activities directly associated with it.

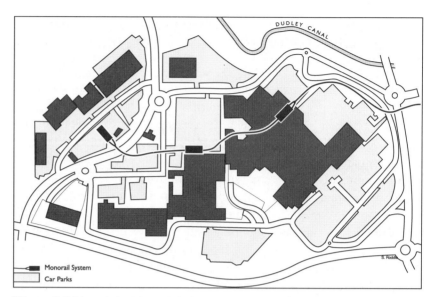

Figure 3.9 Plan of the Merry Hill Shopping Centre, Dudley, West Midlands.

Figure 3.10 Interior of Merry Hill Shopping Centre, Dudley, West Midlands.

It is this intensive specialization that is both a strength and a weakness of such developments. The great shopping street mentioned earlier and many lesser streets in small towns and cities all tend to have one thing in common: in addition to offering retail and recreational services, they embody the spirit of urbanity or *civitas* of their city. It is hard to see such a characteristic ever being generated by a purpose-built commercial shopping centre, no matter now functionally efficient or enjoyable they are.

The 'squares' and 'streets' found in such shopping centres certainly achieve a change of scale, but there is no change in the all-pervasive consumerism of the entire development. These centres are highly efficient and undeniably extremely popular with shoppers, but if their growth results in the demise of traditional town centres, are they replacing all of that which they are destroying?

Perhaps of the four city functions discussed in this chapter (offices, shopping, industry and housing), industry has undergone a more fundamental change than any. From being the prime economic base of most cities through the nineteenth and much of the twentieth century, it has declined rapidly as radically new materials, power sources and production processes have been developed.

Traditionally industry was located in a ring round the city centre, with concentrations in certain key areas depending on local conditions such as the location of canals and railway terminals. Some industries such as shipbuilding, coal mining and textile production in the nineteenth century became inextricably linked with the social culture of the region.

Over the last 50 years, many industries have been run down to such an extent that they play only a small part in the wealth generation of the country. Even the newer replacement industries such as car making and electric appliance manufacturing are diminishing in importance, and the nature of the workplace for almost all members of society is changing. These changes are reflected in the new industrial location patterns that have emerged over recent years[7].

- **Large-scale heavy industry** no longer tends to be based in cities and indeed its relocation away from the major city areas has resulted in much derelict land being left, often severely polluted with the aftermath of over a century of chemical waste spillage.
- **Medium to light industries** can be found throughout the inner rings of industrial cities. They have usually occupied their sites for many years. As others around have either closed down or relocated to an industrial estate or new town, they now find themselves in a complex mixed-use area.

INDUSTRIAL

The morphological characteristic of the urban space in such areas is amorphous with little or no directed thought being given to its quality. Industry is an introverted activity and the streets and spaces surrounding the factory are looked upon almost exclusively as providing service facilities and allowing raw materials to be delivered, finished products to be dispatched and workforce to have access. A city, however, is a working organization and such areas as these are as inevitable and possibly as essential to the economy of the city as those districts more in the public domain and pleasing to the eye. The aesthetics of unpretentious functionalism must be acknowledged as having as much validity in planning terms as public squares and boulevards.

There are three other types of industrial areas we must look at.

- **Industrial estates** exemplified the new policy for zoning and inevitably created single-function areas. Their form is low key with fairly bland, faceless industrial buildings on sites serviced by a road system with grass verges, tree planting, occasional flower beds and large areas of tarmac for parking and vehicular access. The industrial equivalent of housing estates, they were usually located at the outer edge of towns in low-density areas and certainly provided both better conditions for the workers and a better quality environment for their neighbours. They did, however, share a characteristic with the commercial centres in that the entire estate was devoid of human activity at night and during the weekend.

- **Business parks** are a recent development whereby blocks of offices for commercial and other compatible uses (such as computer systems/software manufacture, import/export companies and professional offices) are grouped together in a landscaped area – usually near a major trade centre, a motorway or an airport. The pattern tends to be one of several large office blocks (possibly four or five storeys) surrounded by car parks, access roads and lavishly landscaped areas. Again as in industrial estates and housing, the area is deserted in the evenings and at weekends, leading to 'dead' districts and major security problems.

- **Hi-tech science parks** are a more recent innovation (Figure 3.11). A response to the growth in computer-based sunrise industries, they are often developed as a partnership between local government, enterprise departments and local universities and are adjacent to the universities. They also follow the pattern of buildings in a landscape with access and parking; however, by being located next to a university campus, they can benefit from the activity generated by

Figure 3.11 Aston Science Park, Birmingham, with university main campus in the background.

university life. The pattern of planning is not dissimilar to a campus and therefore the science park can be seen as an extension of the campus both physically and to a certain extent functionally.

Security is usually a high priority feature at such parks and often television monitors provide 24-hour surveillance. This is indicative of a trend also found in the city centre whereby increases in crime and civil disorder are being countered by an increase in the use of sophisticated public surveillance systems.

In summary, the four land-use categories we have discussed so far (commercial, housing, retail and industrial) have collectively and individually, through their own internal evolutionary changes, stamped their characters on the spatial patterns in our modern city.

What are the effects of this distribution on the nature and quality of this space? We will address this question by using four categories of spatial change: accretive, infill, nodal and sectoral zones.

● **Accretive development** occurs primarily at the edge of the city and tends to reflect the general low density found there where housing, suburban shopping, industrial estates and business parks exploit the lower cost of land and where car ownership is rising. This allows developers to feel few inhibitions with regard to access and mobility when locating their development. They are unlikely to provide any public space, as such space presents the private developer with management and maintenance problems. Streets are bordered by minimal front gardens, surrounded by hedges and fenced in, and there is little evidence of real usable external space for the public.

- **Infill developments** are by their very nature more influenced by their surroundings. As sites become available, they often provide the local planning authority with an opportunity to create public space, but such space is expensive as it must be valued in terms of what alternative development the site could have sustained. Many cities have exploited infill sites to remodel the areas by:
  - zoning the area for a new use, such as public open space;
  - carrying out an urban project themselves to their own brief;
  - preparing a planning brief to issue to private developers.

These methods, especially the last, are often used to stimulate and encourage development in inner-city infill sites as developers tend to prefer the cheaper land, usually with less restriction, which can be found at the edge of the city.

- **Urban nodal growth** often emulates the centre: there are small subcentres with pedestrianization schemes for the shopping area and office blocks to accommodate local commercial needs. Even housing is built at increased density, often three or four storeys, and the entire area is a curious mix of urban and suburban characteristics.

  The spatial qualities reflect the local uses with multiple small shops apart from the usual supermarket and an absence of 'civic' buildings. Occasionally, if the subcentre was a historic village, the traditional grouping of church, pub and village green can be found adjacent to the commercial and shopping centre.

- **Central city**: the modern city centre now has commercial viability as its dominant goal. For many years, the dogma demanded that primacy of access and mobility should be given to the motor car with vehicular mobility in the central area making the pedestrian subordinate. There was also little attempt to create high-quality space and it was not until the following series of events started that the pattern changed:
  - Large department stores started to move out of the older town centres.

- Downtown malls were developed by private money and they created private, climate-controlled space which opened only during shopping hours.
- Large out-of-town shopping malls opened up and started to win customers from the older town.
- There was severe criticism from pedestrians about the lack of pleasure in being in the older town centre with traffic conflict, poor road surfaces and danger from crime being cited as main causes.

Most cities firstly recognized the need to conserve and to refurbish their best buildings and then, perhaps for the first time, realized that the pattern of pedestrian movement in town had to be considered as a sequence of streets and squares that linked malls with shopping streets, civic squares and entertainment areas. Eventually there began a movement to exploit historically significant attributes such as docklands, canals, etc., to attract the tourist. Perhaps tourism, the fastest growing industry in the world, is the greatest force for change in urban planning and management today.

WORKPIECE 3.3

CONTRASTING DOWNTOWN AND OUT-OF-TOWN DEVELOPMENT PATTERNS

(a) Select a town centre known to you and analyse its land uses in terms of:

- retailing (on streets or in malls);
- office blocks;
- public buildings;
- social and recreation buildings.

(b) Choose an out-of-town shopping centre and carry out a similar analysis to that specified above.

Now compare (a) and (b), being careful to note the different roles the two play in the urban system. Speculate as to what each has to offer the conurbation in terms of functions and services and the relative merits of each as a spatial experience.

People travel much further for their holidays now and soon realize what city centres could be when they go to other countries. This has made them much more critical of poor environments and has raised levels of expectation as to what could be.

An example of sequenced spaces and squares is found in Milan (Figure 3.12), where the Piazza del Duoma enters the Galleria Victor Emanuel, which leads through a cross-shaped plan to the Pallazzo Marina and then across to the Piazza della Scala. These squares and gallerias combine religious functions (the Cathedral) with shopping (Galleria), and

LEARNING FROM OTHER PLACES

77

Figure 3.12 The Galleria in Milan – constantly full of people all day, every day and open all the time.

civic (Palazzo Marina) with entertainment (Scala), in a composition that creates a true sense of urban presence – truly places, not spaces[8].

This integration of space – linear/collective, civic/commercial – into sequential experiences utilizing urban elements only found in town centres would seem at least the start of the answer to the great out-of-town malls.

THE CITY SQUARE

There is little doubt that, throughout history, the growth and development of the city square has to some extent epitomized the true quality of urbanity found in many of the world's great cities. From the Greek Agora to the Roman Forum, the mediaeval market square and the Renaissance piazza, they all gave a focus to city life and provided a stage upon which history has been acted out over time. Very few of these squares were conceived as a single design statement; in fact many have taken centuries to evolve into the form which we now see. What they all have in common is a scale and quality of environment which, in the words of Aldo von Eycks, create a 'place' and a sense of space. They often share another common characteristic which is that they do not necessarily rely on the quality of their architecture for their effect.

Perhaps one of the most famous is the Piazza San Marco in Venice (Figure 3.13). This square started out as a muddy market site in the

Figure 3.13 Venice, Piazza San Marco.

forecourt of the eleventh century Basilica St Marks, the chapel of the Doge. In the twelfth century, the Doge built his palace adjacent to it. Across the Piazza in the sixteenth century, Sansovino built the Libreria Vecchia followed by the Procuritti Vecchio and the Procuritti Nuovo a little later. All these buildings were constantly being remodelled with new elevations and it was not until the nineteenth century that Napoleon built the final building to enclose the space, appropriately called the Napoleon. It had taken 800 years to create the Piazza San Marco that we now know. It was created as the open-air living-room for a densely overcrowded population compressed on to a small island and was designed to represent the dignity and pride of a city state which, due predominantly to the spice trade, had become one of the richest in Europe.

The piazza is dominated by the ornate, idiosyncratic façade of St Mark's and the crucially located but somewhat austere campanile. The other buildings provide a strong, consistently scaled enclosure which, when it turns the corner hinge of the campanile, opens out unexpectedly into the piazetta and the lido. The square is the spirit of Venice manifest, both past and present, and perhaps represents better than almost any other experience the principles that Kevin Lynch sought to identify in his book *What Time is this Place?*.

Another square, equally successful but in a totally different way, is the Piazza del Duomo in Milan (Figure 3.14). This square sits at the heart of Milan's radial plan and is dominated by its vast cathedral. It was started in the fourteenth century, but the elevation to the Piazza was not completed until the nineteenth century. The remaining buildings around the square are of little architectural merit, but it is on the north side where the great Galleria arcade/shopping street opens on to it that the true innovation of the piazza can be seen.

The Galleria Victor Emanuel not only gives people a covered walkway and gathering point but it also connects Piazza del Duomo to Piazza Scala. This second piazza contains the famous opera house and a magnificent palazzo – the Palazzo Marina, now the city hall. This composition of places, buildings and events creates a dynamic promenade and can be seen being enjoyed by hundreds of Milanese and tourists every day of the year.

The Piazza del Duomo is an altogether less self-conscious space than the Piazza San Marco, yet it too reflects the character of the city: noisy, hustling, commercial but at all times vital and lively. It forms part of a sequence of places, each of a different character, whose total is infinitely greater than the sum of its parts.

Perhaps the most unlikely square of all is Times Square in New York. It provides living evidence that quality of environment is not an

Figure 3.14 Milan, Piazza del Duomo.

essential component for a successful square. For a start Times Square is not, under any criteria, a square. It is an intersection of four or possibly six streets, with five being on the standard Manhattan Grid. It is the remaining street, Broadway, running at a diagonal to the others, that creates the space which has been adopted by the people as a square. The architecture has little to commend it and is constantly changing due to development. The resident activities are seedy and occasionally bizarre, yet this square more than any other encapsulates the spirit of not only New York but the entire nation.

Times Square proves a very confusing and occasionally disturbing experience for an urban designer seeking explanations for its success. It appears to break all the rules and relates to no square performing a similar role in any other country. Is it unique? Yes, of course it is; however, the message it gives us is not. Places are made as much by people by designers and builders.

It represents a whole group of squares whose social, political and ceremonial functions have become more important than their urban form or architectural quality: Place de la Concorde (Paris), Tiananmen Square (Beijing), Wenceslas Square (Prague), Red Square (Moscow) and many others.

WORKPIECE 3.4

EXPLORING CHANGES IN PUBLIC PLACES

Select a space or square in a town you know well and consider its function and whether this has changed through time.

    Identify any changing functions and activities. For example:

- social (informal and formal);
- commercial;
- cultural or spiritual.

Examine also how space relates to its surroundings, particularly the streets and paths which lead to it.

SUMMARY

Until the twentieth century the morphology of a city was an integrated relationship between buildings, streets, squares and open spaces. The history of a city could be deduced from this pattern: the narrow 'romantic' streets of the Middle Ages, the geometrically precise squares and piazzas of the Renaissance, the grand squares and boulevards of the nineteenth century Baroque, all creating an unmistakable awareness of history which is enriching the present.

The twentieth century has led us to create a division between the patterns of buildings and the pattern of streets. Large-scale road systems designed to allow much faster movement through cities led to this separation, with buildings being planned on a different geometry to that of the street pattern. These buildings were serviced by their own sets of access roads taking traffic from the main roads to their point of entrance. Thus the coherent morphological patterns of previous centuries were destroyed.

Each period in history generated a pattern of spatial morphology which reflected its contemporary functions and values and the collage of these urban forms has added to the diversity and dynamic of the modern city's experience [8,9]. This drift away from the morphological cohesion of the pre-twentieth century to the present model of 'buildings in space' may signify more than a new development in urban form.

Firstly, it represents the supremacy of vehicular over pedestrian movement. Secondly, it created an environment in which more undeveloped land was required mainly for car parking, and this contributed to the losss of enclosure which existed when the buildings and streets shared the same geometrical layout. Thirdly, it made little contribution to the characteristic urban forms and patterns of pre-twentieth century cities. The clarity of spatial morphology through historical collages and accretion is a quality greatly enjoyed by city dwellers.

There is, however, evidence to suggest that the socio-economic and cultural structure is no longer clear, unified or harmonious. A better and more accurate description might be erratic, highly strung, pluralistic and culturally diffuse and this must make us question our traditionally held views as to how contemporary urban form should reflect the society which produces it. It may now require a shift of lateral thinking on our behalf in order to interpret accurately what is happening in our present multi-function, pluralistic cities. As with many theories, the action has probably taken place and we are simply seeking an explanation of what has happened and why.

CHECKLIST

Upon completion of this chapter, you should have a greater understanding of:

- the relationship between urban space and the various major functions and activities to be found in the modern, post-industrial city – housing, shopping, offices and industry;

- how major elements such as roads either integrate or divide communities through the effectiveness or otherwise of the public space they offer;

- the distribution of urban space through the modern city plan and the spatial morphology it creates;

● what qualities of experience good, well-organized open spaces in city centres can provide for the citizens.

REFERENCES

1. Heckscher, A. (1977) *Open Spaces, The Life of American Cities*, Harper and Row, New York.
2. Bacon, E. (1976) *Design of Cities*, Penguin, New York.
3. Lean, W. and Goodall, B. (1966) *Aspects of Land Economics*, Estates Gazette, London.
4. Chapin, F. (1965) *Urban Land Use Planning*, University of Illinois Press, Urbana.
5. Gehl, J. (1980) *Life Between Buildings*, Van Nostrand Reinhold, New York.
6. Jacobs, J. (1961) *The Death and Life of Great American Cities*, Random House, New York.
7. Kivell, P. (1993) *Land and the City*, Routledge, London.
8. Kostof, S. (1991) *The City Shaped*, Thames and Hudson, London.
9. Kostof, S. (1992) *The City Assembled*, Thames and Hudson, London.

FURTHER READING

Barnett, J. (1987) *The Elusive City*, The Herbert Press, London.
Jones, E. (1990) *Metropolis*, Oxford University Press, Oxford.
Le Corbusier (1929) *Urbanism*, Editions Cre, Paris. Translated (1947, 1991) as *The City of Tomorrow*, Architectural Press, London.
Lynch, K. (1960) *The Image of the City*, MIT Press, Cambridge, MA.
Lynch, K. (1972) *What Time is this Place?*, MIT Press, Cambridge, MA.
Mumford, L. (1961) *The City in History*, Penguin, London.
Relph, E. (1987) *The Modern Urban Landscape*, Croom Helm, London.
Sitte, C. (1947) *City Planning According to Artistic Principles*, Phaidon Press, London. (First published in Vienna, 1889.)
Sudjic, D. (1992) *The 100 Mile City*, Deutsch, London.
Webb, M. (1990) *The City Square*, Thames and Hudson, London.

THE QUALITIES OF PLACES

PART TWO

EQUITY AND ACCESS

DAVID CHAPMAN AND JOHN DONOVAN

Are there fundamental factors which affect the ways in which we can use and enjoy places? Do some places limit the ability of different people to meet their needs from their surroundings? What attitudes and values affect the decisions which shape the built environment? This chapter looks at some of the ways in which built environments distribute benefits between their occupants and users, and explores the relationship between equitability, sustainability and accessibility.

After reading this chapter you should be able to:

● identify the nature of environmental equity and welfare;

● consider the ways in which the built environment generates and distributes environmental welfare;

● explore ways in which public policies and urban design can foster more sustainable development processes and patterns, ameliorate inequities and increase access for all sections of society.

Towns and cities are the setting for the lives of a growing majority of the world's citizens. They provide cultural and emotional as well as physical fulfilment and often possess great qualities of place and identity. They are also great consumers of physical and human resources, creating wealth and waste, and access to the benefits which settlements can offer may be unequally distributed. Competition for resources and the polarization of wealth often produce disparities and disadvantage between people and places.

Although public policies are frequently designed to redress the balances, they often fail to do so or ameliorate only the worst disparities. Economic disadvantage and poverty may be compounded by poor access to facilities and services which can be denied by the physical form of places, inadequate transport and the inappropriate distribution of services and resources. The level of access can be restricted and this may differ with age, gender, health and economic circumstances.

This chapter explores the concept of environmental justice and the ways in which built environments create and distribute environmental welfare. It addresses these issues in relation to sustainable social and environmental development and the characteristics of equitable urban places.

UNDERSTANDING PEOPLE'S NEEDS

Although everyone in the world has virtually the same physiological needs (for air, water, food and shelter), every individual has distinct personal aspirations and wishes.

There are also quite different values and attitudes to be found in different societies and localities and within dispersed communities of interest. These are sometimes found in national and regional situations but, even in ostensibly homogeneous groups, there is a great diversity of personal needs and wishes.

The built environment plays an important role in enabling people's needs to be met in physical terms.

It can also present obstacles. The concept of physical or environmental determinism, or the idea that the physical form of places can condition or control individual behaviour, does not bear close examination. There is little or no evidence that people will all behave in a similar way if placed in the same physical environment, but it is clear that some physical forms do limit our freedom of action. The most obvious example of this is a prison where inmates are held and their movements are restricted and controlled. Even though there is no direct connection between the physical environment and people's behaviour or reactions, there is a probability that some conditions could make certain reactions and actions more likely.

How does this affect people in the built environment? Are there forms of development that encourage freedom and activity (Figure 4.1), and are there others which inhibit people and the interactions between them?

Figure 4.1 Bury St Edmunds, a thriving and bustling market town.

The form of the built environment as we experience it exerts important influences upon our enjoyment, range of experiences and, importantly, our opportunities. These influences include the distribution of facilities, the means of gaining access to facilities and the nature of the experiences created by the physical form and environments.

The formation of the built environment consumes resources – both physical materials and human labour – as well as creating capital assets. The consumption of resources may have immediate financial costs but may also have longer-term implications for future supplies of materials and the environmental fortunes of the earth.

The distribution of facilities and services and their quality are inextricably linked with environmental welfare. They cannot be divorced from the broader issues of civil liberties and personal freedoms and we must remember that even the most beautiful, accessible and lively place cannot be enjoyed or contribute to its occupants' welfare if they are disadvantaged or cannot gain access to its benefits. The distribution of environmental welfare is not only a matter for the present. It is affected by actions from the past as, indeed, future welfare will be affected by actions of today.

There is now a broad consensus that there are massive challenges for the future of human life on earth. The rate at which we are consuming resources does not appear to be able to continue or to be sustained. The rate of resource consumption is also rising as the developed world produces increasingly sophisticated and diverse products and the develop-

**INFLUENCES OF THE
BUILT ENVIRONMENT**

**DISTRIBUTION AND
WELFARE**

SUSTAINABILITY

ing world becomes increasingly able to desire and demand similar products and access to resources.

The concept of sustainability has become central to the debate about these issues in the last few years, but what is it? The simplicity of the term belies the complexity of the subject, which raises questions about people's needs and how resources are distributed, today and in the future. It is being considered at both national and international levels by politicians and advisers. The discussion deals as much with intra-generational issues of fairness as intergenerational ones. McIntosh offers the following elegant definition:

> Sustainability is a simple concept: living with each other within the means of nature. We must reduce our resource consumption and waste production, and we must make sure that everyone can live decently[1].

AGENDA 21

The Rio Earth Summit in 1992 testified to the importance of the issues at stake, drawing world leaders together to seek international agreements on positive steps which could be taken and measured to reduce resource consumption, depletion and environmental pollution. One of the outputs of Rio was the agreement under 'Agenda 21' that the integration of land use and planning, energy conservation, waste management and a variety of other issues would be examined at local level in consultation with local people. This agreement emphasizes that achieving sustainability depends on the contribution of local people and 'traditional knowledge, values, life experiences or place in a broader society or culture' [2].

SUSTAINABLE DEVELOPMENT

A commonly quoted and useful definition of sustainable development is provided by the Bruntland Report of the World Commission on Environment and Development as 'development that meets the needs of the present without compromising the ability of future generations to meet their own needs' [2].

The objective of equitable sustainability is the maintenance of the quality of life within the carrying capacity of the planet. 'The challenge is to design and manage human settlements in such a way that all the world's people may live at a decent standard based on sustainable principles' [3]. This challenge requires us to find ways of utilizing the qualities of our environment effectively to ensure that everyone has access to the qualities from which they can derive environmental welfare and that those qualities are still available for future generations. This means that we should seek to understand the implications of the effects of our actions on the ability of the environment to generate and distribute environmental welfare. Questions of environmental justice are central to sustainability, and if environments are to be truly equitable they must also be sustainable and continue to meet people's needs into the future.

In their book *Reviving the City*, Elkin and McLaren [4] offer four principles for sustainable urban development: futurity, environment, equity and participation.

FUTURITY This is the concern for the impact of actions of the present upon the opportunities and welfare of people in the future. Though an apparently simple principle, this raises many issues. How far into the future will our action have effects? Are we able to assess them accurately? Can we take tough decisions if desirable actions for the future require us to make sacrifices today?

One of the most significant known effects of current activities on the future is the use of radioactive materials, whose harmful effects can last for centuries. There are also potentially more insidious and devastating consequences from global warming or the extinction of species.

ENVIRONMENT It is evident that our environment is directly and indirectly affected by the policies we adopt and activities in which we engage. By consuming renewable and non-renewable resources, we could deplete the 'stock' of the earth's 'capital'. Pollution is created by waste and emissions and, though the environment can absorb a great deal, there is a real danger in the increasing rate of pollution. How do we measure the full costs of our activities in the long term? Are there approaches which could be adopted that will minimize future problems?

EQUITY Certain issues of intra- and intergenerational equity are important for future relations between the developed world and the developing world. How can developing countries be asked to limit their resource and energy consumption today for the benefit of future generations, if the developed world cannot or will not reduce its own?

PARTICIPATION Policies and programmes are likely to fail unless they enjoy widespread support and public ownership. This can only ultimately be achieved if people are really involved in decision making and ongoing management. The processes needed to facilitate and enable participation are not easy or even comfortable. Many places still suffer repressive regimes and, even in democracies, local involvement is difficult to achieve.

BUILDING FOR SUSTAINABILITY

Brenda and Robert Vale have suggested six principles for planning and designing more sustainable buildings and environments [5]. The choice of the term 'more' is deliberate. As other authors have pointed out, we cannot be sure about the consequences or outcomes of our actions. Scientific measurement may be difficult or ultimate consequences unpredictable. Perhaps the best we can do is to adopt principles which minimize potential future harm. Two potential approaches are suggested.

- **The probability principle**: decisions and actions based upon the probability of beneficial outcomes or responses.
- **The precautionary principle**: policies and methods which offer the prospect of minimization or reduction of harmful impacts, especially where scientific evidence of future impacts is not absolute [4].

The Vales' six principles [5] are summarized below and together provide a useful checklist when we are considering future decisions and designs.

CONSERVING ENERGY   Buildings use energy for a variety of purposes, particularly heating, cooling and ventilation. Buildings can, however, be constructed to maintain thermal comfort levels and be ventilated using natural heating, cooling and air movement. If we do this effectively the energy requirements to supplement these natural systems will be minimized. Thermal mass and good insulation can resist solar heat gain and heat loss, respectively. Management of energy-using systems can minimize situations where, for instance, energy is used to cool or extract air which was previously heated artificially. Expertise and capability in designing buildings and using artificial intelligence systems in this effort is developing rapidly.

WORKING WITH CLIMATE   Each part of the world has a distinctive climate and microclimate. Each offers different potential for utilizing the climatic characteristics to modify comfort conditions within and around buildings. The most obvious example is the use of solar heat gain to warm buildings in colder climates but also in warmer ones in winter. In hotter climates, tall narrow streets give shade from the heat and induce natural air flows to cool them.

MINIMIZING NEW RESOURCES   This principle can be seen in two ways. The first is the minimization of use of non-renewable materials in new buildings and seeking to make sure that the resources used can be re-used in the future if the building or structure is no longer needed. Secondly, existing buildings represent a huge expenditure in natural resources and could possibly be adapted and re-used with minimal resource depletion. In every situation where people understand that resources are limited, both would be natural reactions. It is possible, however, that new techniques or ideas of settlement form (such as the 'compact city') could at some time suggest widespread replacement of existing buildings, raising other questions about the conservation of buildings for their historic associations.

RESPECT FOR USERS   Users of buildings and places are not inert commodities. They are affected by, and affect, the conditions of the place they occupy. They may also be involved in the creation and management of the place. The ways in which they interact with their setting should be respected and understood, and they should be involved in the processes of changing and managing that setting.

RESPECT FOR SITE   Every building operation modifies the condition of the site upon which it takes place. Ideally the impact should be minimal and beneficial. Topography, hydrology, ground conditions and ecology can all be affected both deliberately and inadvertently. An example of deliberate action which can have undesirable consequences is the drainage of surface water. In large developments the speed of runoff from roofs and paved surfaces can cause local drains and water courses to overflow. This is because the natural absorption of the earth has been changed. There are ways of designing permeable surfaces and holding ponds which allow absorption and slows the rate of runoff.

HOLISM   Each of the principles described above is valuable in its own right. Together they provide a robust and coherent approach to building sustainable buildings and places. In each situation we can seek the best balance of advantage and the minimum use of resources.

The very process of building involves the use or appropriation of environmental resources. Construction and use of the built environment consumes resources and also affects their distribution. We cannot be sure about all of the implications and effects of our actions on the built environment and its long-term robustness, but it is important that we do consider the probable outcomes and seek solutions that will minimize long-term harm or risk.

WORKPIECE 4.1

ACHIEVING MORE SUSTAINABLE PLACES

There are a great variety of things which could be done to make places more sustainable and less environmentally damaging.

Consider what they are and who could be involved in each activity. What opportunities or obstacles might there be for collaboration and action?

EXAMPLE 4.1

THE ECO HOUSE, LEICESTER (WRITTEN BY MAURICE INGRAM)

In 1990 Leicester was selected by the Royal Society of Nature Conservation and the Civic Trust to become Britain's first Environment City. It has since received international recognition from the United Nations as one of the 12 best environment initiatives in the world and has been selected by the European Community as a scheme, with a grant of £950 000.

The aim of Environment City is to make Leicester a model of environmental excellence. This involves translating the concept of sustainable development into city-wide action – making it understandable, putting it in place on a local scale and setting up practical projects to show environmentally friendly alternatives to our current life styles.

One such practical project is the Eco House, an environmentally friendly show house set up and run by Environ, the largest local environmental charity in Europe.

A key feature of the Eco House is that it does not look very different from a typical 1930s detached house: in fact that is what it is, but renovated and adapted to include hundreds of features that everyone can include in their own homes, usually simply and cheaply, to make them more environmentally sensitive.

The most important theme at the Eco House is energy conservation. The use and misuse of energy is possibly the single most important environmental issue today. It is also the easiest for homeowners to address.

We use energy in our homes for heating, hot water, lighting, cooling and for running electrical appliances. Environ estimates that, in Leicester, the average house built more than 20 years ago could use at least 30% less energy if only basic improvements were made.

Another theme of the Eco House is to promote ecological building design. The main principles are to:

- maximize solar gains;
- ensure living spaces (living rooms and bedrooms) face south;
- build to high insulation standards;
- design for low running costs;
- utilize wherever possible recycled/reclaimed materials;
- avoid synthetic toxic materials;
- specify energy-efficient appliances;
- design for good ventilation.

A third theme of the Eco House is the Conservation Garden. A garden is a place to play and relax in but it can also offer an opportunity to make real improvements to the local environment. The most sustainable systems in nature are those that have a wide diversity of species and habitats, and these also tend to be the most productive in the long term. The Eco House garden attempts to translate these basic ecological principles into a small domestic garden.

MAKING BUILDINGS MORE SUSTAINABLE

Make checklists of factors to consider, and steps which could be taken, to make:

- a new building environmentally friendly;
- an existing building more environmentally friendly.

We all depend upon our surroundings to provide the facilities and settings within which we can meet our day-to-day needs. These settings include the dwellings that offer us shelter, the fields and gardens that provide food and recreation, the shops and markets that supply provisions, and the public spaces that provide vital links between them and opportunities for social interaction.

Different places have different qualities and can fulfil different needs. All places have the ability to meet more than one need. For instance, a park may provide opportunities to relax, play games or watch wildlife. No space or building is likely to meet everyone's needs. If it did so the occupants, wanting for nothing, would be unlikely to leave. To offer the occupants of any given location the wide variety of experiences needed to meet their needs, then those occupants must be able to access the appropriate settings for those experiences.

When development takes place it changes the characteristics of the site and, by extension, the environment within which it occurs. In changing these characteristics, development affects the range of experiences that are available within that environment. The greater the range of opportunities for interaction an environment can provide, and the more appropriate they are to the needs of the people who use them, then the greater the ability of that environment to meet diverse needs.

The environmental welfare threshold could be described as the smallest area within which all the day-to-day needs valued by the occupants of a place can be met. Examples 4.2 and 4.3 show two quite different outcomes from intervention and development in the built environment.

While these case studies do not compare like with like, they do illustrate the significant differences of impact on access and equity which can occur between different sorts of development. The first requires the people affected to go further afield to find the range of experiences which they previously enjoyed from their environment, whether it be travelling further to find quiet open space, safe attractive

THE BUILT ENVIRONMENT AND ENVIRONMENTAL EQUITY

ENVIRONMENTAL WELFARE THRESHOLDS

play facilities for children, or in order to meet former neighbours. A friend may live only 500 metres away but, if there is a motorway between these two people, this physical barrier can separate them as decisively as would a much greater distance.

EXAMPLE 4.2

REDUCING THE ABILITY OF PLACES TO MEET NEEDS: THE M11 EXTENSION IN LONDON

The profound changes that development can make to the characteristics and range of settings available in an area is demonstrated by the M11 motorway extension through east London. Such an environment loses houses and so is less able to provide the settings within which housing needs are met. It becomes a poorer setting to meet social needs (losing social opportunities and community networks) or cognitive and aesthetic needs (because of the loss of evocative or socially valuable buildings, spaces and landscapes and other opportunities to experience nature). On the other hand, such an environment gains the (environmentally dubious) characteristic of enabling vehicles to move through it more quickly.

The qualities and characteristics that were previously valued by those occupants are no longer available. Promoting vehicle movement has been achieved at the cost of compromising the area's ability to provide settings within which other needs are met and environmental welfare derived. Previous residents have become displaced from their familiar territory, and remaining occupants have to live with the new motorway which occupies part of their day-to-day environment. The capacity of the site to 'carry' the stock of resources/qualities required to meet people's needs has become severely diminished through development.

EXAMPLE 4.3

IMPROVING THE ABILITY OF THE ENVIRONMENT TO MEET NEEDS: BITTON, KINGSWOOD

An example of intervention which is designed to assist the satisfaction of environmental needs is provided by the development of a site for housing in the English village of Bitton, near Bristol. Apart from any other requirements, the guidance stressed the need for an element of public open space and for visual and physical access to surrounding countryside, as well as outlining some of the characteristics required of these qualities. In developing in accordance with these objectives, apart from removing an eyesore the existing and incoming residents will gain improved access to the settings for experiences previously difficult or inconvenient to get to (for example, the

surrounding countryside and the village centre) or previously unavailable to them (for example, a pleasant outdoor public place). All these objectives were derived from an understanding of the social landscape of the villages occupants. By introducing into their surroundings qualities that are relevant to the perceived needs of the occupants, the development increased the potential of that environment to enable people to derive environmental welfare from it; consequently the carrying capacity of the site to meet people's needs had been dramatically improved.

In the second example, by contrast, the people affected by the development find that they can travel much shorter distances to meet their needs and to experience the qualities which had previously been difficult to get to or had been unavailable.

What can we conclude from this? In diminishing the range of opportunities, qualities and experiences available to occupants of a place, or making them look further afield for replacement qualities, the environmental welfare threshold is extended. In increasing the range of opportunities, qualities and experiences available to occupants of a place, or saving them from looking further afield for replacement qualities, the environmental welfare threshold is reduced. These thresholds vary from place to place and from time to time. They can be described as the range within which people can travel or move conveniently to meet their needs from their environment.

Increasing the need to travel, or to travel greater distances, has important environmental and equity implications: the greater the distance, the more limited may be the choice of modes and the greater the resource and human costs. Maintaining or improving the level of environmental welfare and reducing the distances which must be travelled to meet needs will reduce the resources required to access those needs in terms of individual time, energy for transport and infrastructure provision, and the space taken up by transport corridors.

A number of important points arise from this and together provide a theoretical basis for the concept of environmental welfare thresholds.

- For any given environment there is a range of experiences and qualities that can be gained from it, and the opportunities and experiences available to individual occupants is directly related to the range they can access.
- Everyone who occupies that environment has a range which they can realistically travel to gain access to the facilities and resources which it offers. This depends upon their individual mobility and thus influences their individual environmental welfare threshold.
- Development of any sort can change the capacity of that environment to meet people's needs and, consequently, affects the environmental welfare threshold. This can be both beneficial, where new qualities are valued by the users of the place, or harmful, where existing qualities are appropriated or severed from those users.

From this it can be seen that where development increases the environmental welfare threshold it reduces welfare by denying access or requiring people to travel further and consume additional resources to access the qualities and facilities they value. The resource implications relate directly to global and local sustainability. As global resources are depleted, the use of energy to provide private and public transport may be diminished. Environments with compact thresholds are likely to provide and maintain welfare even if personal mobility and range are limited.

EXAMPLE 4.4

ESTABLISHING EQUITABLE OBJECTIVES IN BRIDGTOWN, STAFFORDSHIRE

Bridgtown is a small English industrial and residential settlement, home to some 700 residents, which has been identified for industrial development in all post-war plans. This, and the extensive demolition of derelict and unfit housing, has resulted in a dramatic population decline to one third of its 1950 high of 2000. A parallel decline in the services that the community supported, and relied upon, has also been evident. Only one church remains and many shops and pubs have closed down.

The range of interactions that was available in Bridgtown was reduced between people and places as fewer activities could be supported, and amongst people as there were fewer of them around with whom to interact. Against this perceived and unwelcome spiral of decline, a plan was drawn up based on the residents' and visitors' social landscape. The plan was based on a survey which asked people five questions:

- What is good about Bridgtown?
- What is bad?

- What symbolizes Bridgtown for you?
- What are the most important issues facing Bridgtown?
- What would most improve the village?

There was a remarkable degree of agreement between people about the aspects of their neighbourhood that were significant (the social landscape) and their relative importance (the social agenda). The resulting plan has been instrumental in convincing the district council to amend the local plan in favour of residential rather than industrial development and to commit resources to various identified environmental improvements.

The plan can be seen as the synthesis of the community's environmental aspirations, translated into planning terms. As such it was a means of communicating, and eventually achieving, the measures that would better equip the people of Bridgtown to derive environmental welfare out of their surroundings.

WORKPIECE 4.3

COMPILING A SOCIAL AGENDA

Pick an area – preferably the one you live in, or somewhere you know very well. Then answer the following questions:

- What is good about the place?
- What is bad about the place?
- What symbolizes the place for you?
- What goods or services do you need to leave the area for, or to get?

- What do you think are the most important issues facing the area?
- What do you think would most improve the area?
- Will other people's answers concur with you own?
- Consider why other people's responses would be different.

The shape and pattern of development can influence pedestrian movement quite profoundly. The solid wall of a road or rail embankment can completely prevent people from passing across it. A busy road can also inhibit many people from crossing it. Both instances create a severance effect and limit human movement and interaction. Similarly, a well-placed gateway or crossing can facilitate movement, but only where the gate is![6].

Not only can access be denied economically as well as physically; it can also be denied or limited psychologically. The fear of danger in certain places or at different times may limit the freedom of movement and access for the majority of people or for particular groups of people – for example, the young, elderly or infirm. Though it will never be possible to make everywhere perfectly safe for everyone (indeed, some people would argue that to do so would diminish the variety of human experience and the vitality of diversity), it is important to recognize the possible effects of actions upon people's safety or their perception of safety (Figure 4.2).

ACCESS AND SAFETY

Figure 4.2 Subway steps, New Street, Birmingham – a major pedestrian barrier.

Safety from crime or accident, and from the fear of them, is important to everyone. Though some may choose to expose themselves to various sorts of risk by gambling, dangerous sports or hazardous behaviour, it is important that the ordinary occupants of the built environment are not forced into positions of danger or threat. The pattern of development and its detailed design can contribute to the level of actual and perceived safety or of threat. It is important, therefore, to consider how any development proposal will be perceived and used in these terms and whether this understanding may help modify and improve existing places.

In terms of the detailed design of dwellings, the useful concept of defensible space was suggested by Newman, who proposed it as an approach to the structure and relationships of space around dwellings [7]. The importance of the ways in which public, semi-public and private spaces relate to one another is explored strategically and in terms of detailed design to give an increasing sense of 'ownership' and 'control' over the semi-public and private space outside each dwelling.

The layout of streets and footpaths can also affect our perception and use of places. It has been argued by Bentley *et al.* [6] that the permeability of places, or the variety of alternative routes through that place, affects the accessibility and enjoyability of it. The existence of a variety of choices of movement pattern to the users of a place may also increase their perception of their personal safety by allowing them to choose which paths and routes to take, avoiding unwelcome experiences and picking enjoyable ones. This is most readily achieved in places which have a close grain in the development pattern. This close grain could be simplistically defined as closely spaced networks of streets and paths which provide a variety of convenient alternatives for users. The street blocks which these networks of movement serve may also have a close grain of buildings and variety of uses. In these sorts of places the proximity of a variety of mixed uses may provide further diversity of choice and minimize the distances which people need to travel to meet their needs.

MOBILITY AND MODES OF TRANSPORT

Much of the access to facilities and services which people enjoy in the developed world depends upon private and public transport systems. We have seen in Chapter 2 how influential transport and communication systems have been in shaping and accelerating the growth and development patterns of settlements. Indeed the availability to many people of affordable private transport, in the form of the automobile, has had a profound effect upon the distribution of services and facilities in the late twentieth century.

In much of the world this has led to dispersed settlement and development patterns which depend on high levels of motorized mobility. In Chapter 3 we saw some of the forms that the built environment has taken as a result of these trends, which have accelerated continuously since the 1950s. They are described by Deyan Sudjic in *The 100 Mile City* [8] which takes a critical view of urban form and the way people perceive and interpret it.

In 1994 the Royal Commission for Environmental Pollution [9] said that the unrelenting growth of transport has become possibly the greatest environmental threat facing the UK, and one of the greatest obstacles to achieving sustainable development. This threat includes energy consumption, pollution and land take as well as many other effects such as the impact of road building upon surface water runoff, hazards to personal safety and severance between facilities and areas (Figure 4.3).

In 1994 it was stated by Boyle [10] that 'the transport sector consumes around 25% of global primary energy use, while emitting 22% of energy related carbon dioxide emissions and that it continues to be the fastest growing energy sector in many countries'. He went on to point

Histogram of Passenger Kilometre by mode.

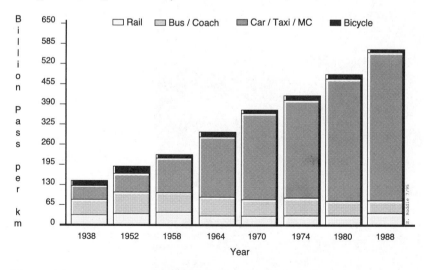

Figure 4.3 The problem: ever increasing traffic. Inland travel in Britain 1938–1988.

out that 'local air pollution will remain a key issue as the benefits of increasing use of catalytic converters and lead free petrol are undermined by the sheer volume of vehicles on our roads thus raising concerns over the health of children, pregnant mothers and the elderly'.

Not only does motorized transport produce pollution, noise and danger but it also creates potential barriers to movement in other ways and by other modes. Roads, railways and canals all act as linear barriers which can only be crossed safely at particular points (Figure 4.4). These crossings may be at 'grade' or ground level – for instance, at road junctions or pedestrian crossings – or by moving over or under by bridges or tunnels. The nature and frequency of these crossing points has a significant impact upon accessibility at the local level. This means that every system to provide mobility in the built environment will also potentially reduce accessibility at the local level. This is quite an obvious conclusion to draw but is it given enough attention in planning new transport infrastructure?

Figure 4.4 Birmingham Inner Ring Road – a concrete collar.

Increasing the distance between resources and facilities may mean that they are practically accessible only by private car. Those who cannot drive or do not have the use of a car are effectively excluded from many of the benefits of their environment. This includes 66% of women and 30% of men in the UK. To this can be added everyone under 17 years of age and many elderly people who are no longer are able to drive. It should not be forgotten that the car does expand the mobility and horizons of many people, especially those people with physical disabilities. Not only are people able to travel further to access facilities and benefit from opportunities, but those physically impaired can often reach facilities which would otherwise be out of their reach. Specially adapted cars, community transport and shop mobility schemes all transform mobility and individual access.

Many people and groups suggest that walking and cycling offer convenience and attraction, especially if suitable safe routes are available. Here there is an opportunity and a problem. The existing road systems of most places would be ideal for walking and cycling if it was not for the hazard and pollution of motor vehicles. Taylor has proposed the concept of a 'kindly circle' in which the provision of safer cycling and walking routes would reduce dependence upon unsustainable motor vehicles thus making existing roads more attractive for increased cycling and walking [11].

Moving between any two points involves potential risks. The distance travelled and the speed of travel may also increase the chances and seriousness of accidents. Reducing distances and increasing the relative attractiveness of walking and cycling may save life and improve its quality by promoting healthier and less environmentally intrusive alternatives.

One of the most profound changes in human capability over the last century is the capacity which we now have to move people and goods in great quantities, over great distances. This ability increases personal freedom of movement and gives national economies greater equality on the international stage. The liberating qualities of modern modes of transport are not unalloyed. New roads can increase the number of journeys people make, increase atmospheric pollution and create hostile pedestrian environments. Air travel pollutes through noise, exhaust gases and fuel jettisoning, and may dilute the cultural diversity of societies around the globe.

It cannot be denied that much planning and development has been concerned with improving physical access and with improvements in

**ACCESS AND
DEVELOPMENT
PATTERNS**

transport and communications. This has inevitably led to conflicts of interest as one person's mobility may be another person's nuisance. There is a fundamental question here. Can individual access be provided without compromising the access or environmental welfare of others?

SPATIAL DISTRIBUTION – LAND USE AND TRANSPORT

One of the most significant developments in public policies of recent years, following Agenda 21, is the attention now being paid to land use patterns and their effect upon transport and access. It may seem surprising that this has not always been so but until quite recently the idea of locating different uses in separate zones was conventional wisdom. Similarly the idea that market forces would be sufficient to direct the efficient distribution of activities held sway for much of the 1980s and early 1990s.

The results of these policies can be seen in the business parks and office areas deserted after working hours, and the sprawl of out-of-town retail parks. The consequences of separating activities is threefold:

● It necessitates travel between facilities.
● The movement affects the places situated along the line of travel.
● The costs of travel in time and money may affect the ability or frequency of some people to access those facilities.

The concept of integrating land uses and distribution of facilities to minimize the generation of journeys while maintaining and improving access to facilities for everyone is very important. Here we have the potential to increase or maintain individual welfare while containing energy consumption and pollution locally and globally. But can we rely upon this response alone? Will people really be content to stay in their integrated 'urban village'? There is little research to guide us on this but there is evidence that the very process of travel has become a recreation and a freedom which many people enjoy and may continue for its own sake. Conversely the explosion in communication technologies may profoundly change our need to travel at all. But will it change our desire to travel?

ACCESS TO, AND THROUGH, BUILDINGS

As we have seen, the ability to move around is important to accessibility. As many of the facilities and opportunities we seek are frequently located in buildings, it is equally important that those buildings or complexes are fully accessible. This is frequently not the case, especially for people with physical disabilities.

Many buildings, especially large complexes, cover extensive sites. These sites may provide or cover routes through which people wish to pass to gain access to other facilities (Figure 4.5).

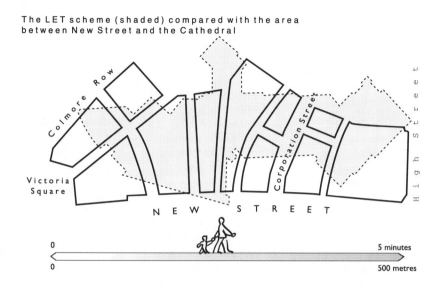

The LET scheme (shaded) compared with the area between New Street and the Cathedral

Figure 4.5 Comparison of urban grain and scale – the plan of a proposed monolithic development shown over the Victorian grid of streets in central Birmingham (J. Holyoak, with permission).

The effect of our decisions at all scales may affect the mobility of some people. The location of facilities, design of buildings and choice of materials can all be of importance. Their impact may affect some more profoundly than others and, in effect, discriminate between people. Careful thought is necessary to achieve built environments which are accessible to everyone.

In other chapters we look at the ways in which people, professions, businesses and politicians interact in shaping the form and qualities of the built and living environment. Without the active involvement of people in problem definition, decision making and implementing plans, projects can be misconceived and misguided. Facilitating true participation is not easy and depends upon the values and attitudes of those involved. It is important that the steps taken to create equitable and accessible environments are inclusive and involve the occupants and users of that place (Example 4.5). Many assumptions – and frequently false assump-

PEOPLE AND PARTICIPATION

tions – are made about what is needed and what people want. Even if the assumptions are sound, it is essential for us all to understand the value of involvement and participation and to recognize the sense of ownership that they can facilitate [12].

WORKPIECE 4.4

STUDYING ACCESSIBILITY OF BUILDINGS

Pick a building that you can get permission to enter. It could be your own house or hall of residence or your college buildings. Using the checklist below, carry out a study of the accessibility of that building.

Then try the same exercise with the needs of different people in mind; for example, an able-bodied person, a mother with a push chair, someone with visual impairment, and someone in a wheelchair.

- Are there any hazards or obstacles outside the building?
- Is the entrance easily accessible?
- Are accesses and facilities clearly signposted?

- Are circulation routes and facilities, especially toilets, accessible to all?
- What characteristics could exclude or obstruct any potential user?

Checklists are only as good as the points they raise. Consider what could be missing from this checklist and try drawing up a list of other points which should be considered.

Remember that different people may have different access needs and that sometimes their individual needs may be in competition.

WORKPIECE 4.5

MOBILITY *V.* ACCESSIBILITY

Find a local area where a new road has been planned, or built, to improve traffic movement, or mobility, through the area. By studying plans of the layout of roads and footpaths before and after construction of the road, try to identify the impact of the road upon vehicular and pedestrian movement, and the local accessibility.

EXAMPLE 4.5

VANCOUVER

The Vancouver cityplan process is interesting because it contains the five characteristics of participation suggested for sustainable communities for local Agenda 21 collaboration. Key points to note are its inclusiveness, the absence of predetermined agendas, the emphasis on mediation between many competing interests and its participatory nature.

Vancouver is an attractive multicultural city with a pleasant climate on Canada's thriving Pacific coast. It is

Canada's largest port and north America's second largest. These factors contribute to the very high growth rate (44 000 people every year). With this growth has come congestion, spiralling costs and strains on the city s infrastructure. Against this background the city launched the cityplan process, notable because it was open-ended, based almost entirely on public participation, and the only predetermined objective being to achieve greater sustainability. The cityplan process invited people to raise the issues that the plan should address. These issues were then considered by city circles which contained members of Vancouver's many diverse communities and anyone who wanted to attend. The officers of the city council arbitrated between their ideas and recorded the findings. These city circles then presented their ideas at a well-publicized travelling 'ideas fair' to the general public who effectively defined the future of their city.

In this way the cityplan process moved from general aims to specific objectives and gathered a lot of information that would be difficult to ignore. The United Nations Association has identified the characteristics of public participation to achieve sustainable plans:

- Go well beyond established groups.
- Be non-adversarial.
- Mediate.
- Have an open-ended process.
- Reflect local interests and priorities.

It can be seen that the cityplan process (because it is inclusive, has no predetermined agenda and emphasizes mediation and participation) has the qualities required to identify sustainable objectives.

SUMMARY

Distributing the opportunities for environmental welfare so that everyone is equally able to benefit from them depends on equipping environments for the whole range of activities required. The ability of an environment to cater for a range of interactions, and consequently have a range of compatible meanings for its occupants, has been described as its robustness [6].

Providing settings within which a range of interactions can occur will only contribute to the equitability of the environment if its occupants are physically able to get to the relevant settings for the desired interactions. Ensuring that no one is denied access to the settings from which they derive environmental welfare is consequently of central importance. Equitable built environments minimize the elements and characteristics which limit their occupants' opportunities to use and move through it, irrespective of their particular levels of mobility. The ability of an environment to enable people to access its qualities can be described as accessibility.

Enabling people to get to where they want to be is central to making environments equitable. If getting there exposes them to physiological harm, requires them to take unnecessary risks or gives them a feeling of danger, then the ability of that space to cater for purposeful interactions is compromised. The fear of coming to harm is experienced differently by different groups. Consequently, whatever other qualities an environment may have it will be inequitable if it compromises the ability of sections of the community (women, the aged or ethnic minorities, for example) to derive environmental welfare from their surroundings.

Creating the settings and opportunities that allow us to meet our needs, and making them accessible and safe to get to and use, will increase equitability today. However, it should be remembered that not all of the occupants of a given environment will occupy it at the same time. If those who come after us find that they are less able to meet their needs because of the legacy we leave, we are appropriating the future environmental welfare even though all its present users may be able to meet their needs from their surroundings. Equitable environments will have characteristics and built forms that are sustainable in construction, use and demolition.

This leads us to the conclusion that equitable environments must be robust, accessible, safe and sustainable.

CHECKLIST

The issues covered in this chapter are:

- the ways places generate and distribute environmental welfare amongst their occupants;
- the need to understand how patterns and forms of development affect access, mobility and welfare;
- the inextricable relationship between environmental welfare today and sustainable behaviour for the future;
- the value of robustness and adaptability in buildings and places;
- the relationship between land use and transport and their effects upon accessibility and equity;
- the significance of physical and psychological safety upon the use and enjoyment of the built environment;
- ways of building and developing which reduce resource and energy consumption and reduce long-term risks and consequences;
- the importance of involving users and occupants in the decision-making processes about the management and development of places.

REFERENCES

1. UBC Task Force and the City of Richmond (1993) Plan for healthy and sustainable communities. *Planning in British Columbia News*, Vancouver.
2. The World Commission on Environment and Development (1987) *Our Common Future*, Oxford University Press, Oxford.
3. White, R. and Whitney, J. (1990) *Human Settlement and Sustainable Development; an overview*, University of Toronto, Toronto.
4. Elkin, T. and McLaren, D. (1991) *Reviving the City*, Friends of the Earth, London.
5. Vale, B. and Vale, R. (1991) *Green Architecture: Design for a sustainable future*, Thames and Hudson, London.
6. Bentley, I., Alcock, A., Murrain, P. *et al.* (1985) *Responsive Environments*, Architectural Press, London.

7.   Newman, O. (1972) *Defensible Space: People and Design in the Violent City*, Macmillan, New York.
8.   Sudjic, D. (1992) *The 100 Mile City*, Andre Deutsch, London.
9.   Royal Commission for Environmental Pollution (1994) *Transport and the Environment*, HMSO, London.
10. Boyle, S. (1994) Transport background paper. UNA workshop at Sustainable Development Conference, London.
11. Taylor, M. (1994) Women in the built environment. *Streetwise*, **5**, 2.
12. Blackman, T. (1995) *Urban Policy in Practice*, Routledge, London.

FURTHER READING

Harvey, D. (1989) *The Urban Experience*, Blackwell, Oxford.
Lynch, K. (1981) *Good City Form*, Massachusetts Institute of Technology, Harvard.
Richards, B.(1990) *Transport in Cities*, Architecture Design and Technology Press, London.
Tibbalds, F. (1992) *Making People Friendly Towns: Improving the public environment in towns and cities*, Longman, Harlow.

VARIETY AND VITALITY

MARTIN BRADSHAW

THEME

Towns and cities are much more than bricks and stone. The physical form is only a stage set upon which a multitude of human dramas continually unfold. The variety of activities which a place supports, 24 hours a day, seven days a week, is an indicator of its vitality. Activities may be conflicting or mutually supporting. They may be accessible or exclusive. This chapter explores the characteristics of places which make them lively and responsive to their users' needs. It considers some of the ways in which public policies and private activities have encouraged vitality and the ways in which we can all be responsible for the places we use.

OBJECTIVES

After reading this chapter you should be able to:

● understand the qualities which contribute to the vitality of places;

● explore the ways in which varied built environments are created;

● understand how the vitality and viability of places can be enhanced and maintained;

● examine a range of policies and initiatives which have been taken to promote variety and vitality in built environments and communities.

INTRODUCTION

The sense of variety in the built environment of English cities, towns and villages is virtually limitless. Its presence or absence is what we measure, consciously or otherwise, when we make judgements about the quality of a place and whether, as a consequence, we regard that place as

110

exhibiting vitality. What is it that creates variety and vitality? It is inevitable and perhaps part of the attraction of the topic that judgements will be partly subjective. One may prefer the timeless visual cohesion of Bath, with its elegant curving crescents, to the more dramatic and sometimes brash townscape of nearby Bristol, with its water and its flavour of merchants and manufacture. The important point is that it is possible to look at the factors that have led to the variety and consequent vitality of these two cities being so different and, looking further afield, what it is that enhances vitality and what diminishes it.

We start with the premise that the built environment and its variety is much more than bricks and stone and spaces. We should be in no doubt, however, about the infinite mixture they create which is just as great as that of the English landscape, even though it seems to get less literary attention. 'Townscapes', like landscapes, can be wild or tame, strange or serene, lonely or busy, sinister or homely, spacious or confined, sometimes all within the same town. There are towns which are harmonious and gentle, and others which are the result of violent collisions of buildings with nature or each other'[1]. The physical form, however, is only a stage-set upon which a multitude of human dramas continually unfold. Variety, therefore, is not just in the form, but in the activities which a place supports or stimulates, day in and day out, and which are one indication of its vitality.

Activities can be random or organized, conflicting or supporting, accessible or exclusive, but they are almost all a response to the characteristics and attractions of a place.

The concept of attractions is very important and is returned to below, but two other general points need emphasizing first: physical variety at some scale is almost certainly a prerequisite for vitality, but variety cannot guarantee stability. Centres of towns, for example, which have little to offer except shopping will exhibit activity only at certain times. Today, they are increasingly vulnerable to out-of-town pressures, to down-market movement and to a loss of variety and vitality, unless they can respond by broadening their mix and function, which helps to animate their spaces after the shoppers have gone. The second and complementary point is that unless the scale is small, there is virtually nowhere that does not have at least some variety and some elements that give distinctiveness. A choice is still available, therefore, to develop consciously a strategy that uses those features to create greater vitality. This thinking is at the heart of much contemporary urban regeneration and is touched on later in the case studies.

WHAT ARE VARIETY AND VITALITY?

How can variety and vitality be identified and measured? The former may be defined as a combination of townscape scale and qualities, mix and range of uses and functions, and the commercial, cultural and leisure activities that are generated. The measure of vitality is much more complicated and open to misrepresentation. As the Department of the Environment put it in 1992, 'For experienced traders and surveyors, the existence of vitality (and viability) is easy to recognize, but difficult to define'[2].

The word itself implies liveliness and animation and, therefore, might be applied to a town centre in terms of how busy it is. The revised government guidance has certainly followed this line with one of the two indicators it defines as helpful in the assessment of proposals for retail development. This is the measure known as **footfall** or pedestrian flow, which involves the counting of people passing a particular point at a particular time. When undertaken at different locations and over a period of time, it is at least an indicator of how lively a street or centre feels. Over time, it also helps to assess the impact of out-of-town centres as at Newcastle, where the impact of the Gateshead Metro Centre was shown through counts over an 11-year period to have affected the centre as a whole, but not the higher quality Eldon Square development. Counts have also proved to be important in judging the effect of pedestrianization, an increasingly critical assessment, as it becomes clearer that solving 'the problem of unsuccessful public places by eliminating traffic, is not the answer'[3].

The second Department of Environment indicator is **yield**, and this is perhaps applicable more to judgements of viability than vitality, which are not the same. Technically, yield is the ratio of rental income to capital value so that, confusingly, a higher yield is an indication of concern or greater risk, whilst a lower yield indicates attraction. Thus the yield in the city of Sheffield in 1988, before the opening of Meadow Hall, was calculated to be 4.75% but 'fell' to 7.5% in 1992 after the opening[4].

Neither of these indicators provide wholly satisfying measures of variety and vitality, which are concepts that do not simply involve economics but are socio-cultural as well. The comment that 'if only there were fewer tourists, the place would be so much more attractive' reflects the point that attraction for one is not necessarily for all. Nevertheless, a more rounded basis is needed for assessing the qualities that create vitality, and this is looked at in the next section.

Our built environments are complex in character – the product of cultural, social and economic change, of historical events, and of contemporary decisions and pressures. They are, in short, evolutionary. Their variety, therefore, may come significantly from the past, but their vitality depends more on how they are adapting to current trends.

These changes are not confined wholly to town and city centres which we conventionally associate with vitality. Collonge La Rouge, for example, one of France's most beautiful villages, is an historical essay in red sandstone. This village's vitality comes from its small specialist shops and restaurants and its high visitor attraction. Maintaining its environmental qualities is perhaps all that is needed. But town and city centres generally perform a more complex and organic mix of functions as shopping centre and market, business centre and transport hub, as entertainment and leisure venue, a meeting place, a service centre for our education needs and, for some, a place to visit.

The variety and vitality of such centres comes first and foremost from the way in which they respond to the trends and pressures which bear on their multifunctional nature. Some of these may be briefly identified.

DE-URBANIZATION   The trend to dispersal of both people and jobs away from traditional centres has been a feature of change in the UK for 25 years and has weakened vitality, in marked contrast to the reurbanization evident elsewhere in Europe. This neglect of our urban resource is one reason why British centres have lost their residential function, now belatedly recognized as a critical contributor to vitality.

MOBILITY   The increased availability of cars has revolutionized mobility, but has reinforced dispersal and diminished the pedestrian environments of town and city centre. Our centres were not built for cars and, where they have been adapted, the ubiquitous inner ring road and multi-storey car park have diminished their vitality rather than sustained it.

SOCIAL CHANGE   The move to a service economy, and particularly the increase in women returning to work, has coincided with more home entertainment and a fundamental retailing change. The new multiples and shopping malls concentrated first in towns, often to the detriment of their variety. But rising car ownership and town centre congestion, plus a lemming-like tendency to imitate the United States, led

QUALITIES CREATING VITALITY

WHAT ARE THE FORCES AT WORK?

first to the movement of supermarkets away from the centre and then to out-of-town shopping on a regional scale with Merry Hill, Metro Centre, Meadow Hall and others. Being virtually town centres in their own right, they have had an impact which is now discernible and measured [5]. But it is the superstores, which now saturate most urban areas and seem set to threaten quite small market towns, that have caused the most community concern.

LEISURE AND TOURISM   Home entertainment and the video have led to damaging losses in cinema, theatre and dance halls, once a mainstay of vitality. For the historic town, with its heritage and its specialist shops, tourism has helped to counter the trend, and even former industrial cities like Bradford have used industrial heritage to remarkable effect. Shopping itself is now a 'leisure' activity for the more affluent, from which the multifunctional nature of the town centre has profited.

CAN VARIETY AND VITALITY BE PROMOTED?

We come back to something called 'attractions' – the things which might counter dispersal and lack of accessibility, and might tempt people to use a place after hours. This means a high standard of environmental attraction with or without a few historical 'anchors', but it also means a rethink of role with perhaps the school and the library becoming (as they are in the United States) very strong community foci.

To assess the prospects of any strategy seriously, however, needs a better measure of what vitality really comprises. Footfall and yield are indicative but no more. By far the most admirable and measured analysis has been made by URBED for the Department of the Environment's report, *Vital and Viable Town Centres: Meeting the Challenge* [4]. The report set out to establish a practical assessment framework as a means of judging the 'health' of a town. Health is an imprecise surrogate for variety and vitality, but one is dealing with subjective concepts where there are, in any case, different views on what constitutes a successful centre.

The report identified three basic qualities that underlie the health of a town centre: attractions, accessibility and amenity. A fourth quality of action is concerned with 'making things happen' where, as in most cases, there are deficiencies in the strength of the first three.

ATTRACTIONS   Attractions are seen as the foundation of a healthy centre, but not solely in terms of the mix of retailing. Arts, cultural or entertainment facilities, education, health or other services, and the availability of space for living and working all help to keep a centre alive. Figure 5.1 indicates some of the relationships in simplified form and highlights the importance of critical mass, or the concentration of enough activity to provide visible choice.

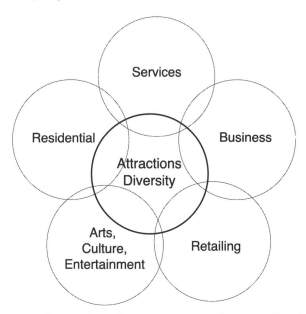

Figure 5.1 Attractions include diverse elements and services (developed and adapted from *Vital and Viable Town Centres: Meeting the Challenge*, HMSO, London).

ACCESSIBILITY **Accessibility** is identified as a measure of 'how easy it is to reach the centre' and, particularly in cities, how easy it is to move about. The DoE report further defines two components. **Mobility** refers to the 'time and cost of getting to the centre from where people live'. It is here that the balance between modes of travel, traffic management and car-parking provision, and between the needs of car-borne travellers and a high standard of amenity, become critical. Until recently, the evidence suggested that the exclusion of the car has been associated with retail success and has enhanced attraction and vitality. Economic uncertainty, however, now faced by many town centres, has made the approach much less certain, particularly for market towns. **Linkages** or local accessibility refers to the integration between transport facilities, and their relation to the main attractions and to ease of movement, navigation and therefore signing: somewhere, perhaps, that makes the geographically dyslexic feel comfortable. Figure 5.2 shows some of these relationships.

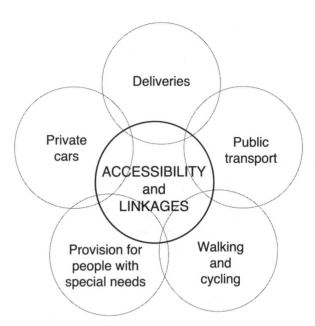

Figure 5.2 Accessibility depends on many factors (developed and adapted from *Vital and Viable Town Centres: Meeting the Challenge*, HMSO, London).

AMENITY   Amenity is the third component of health and vitality, or 'how pleasant a centre is as a place to be in'. This is the most difficult to define. It is obviously a reflection of an attractive and well-maintained environment, but the DoE rightly identifies two other elements:

● **Security**. Good maintenance and cleanliness are reinforced if a place is felt to be secure. Paradoxically, however, security is perceived to be inversely related to the presence of cars, i.e. their exclusion may help in the day, but is threatening at night.

● **Identity**. The tendency for too many centres to look the same and lose distinctiveness is growing. It is partly due to retail concentration, loss of small shops and concern for corporate image. The number of individual shops has halved in 30 years, whilst floorspace has actually increased by some 30%. The leading chains of multiples are everywhere and make few concessions to local character. But loss of identity comes also from the failure to protect heritage, from poor public space, from too much mediocre new development and the loss of familiar things. As the report points out, 'the experience most people have of a town is largely based on the smaller things, and attention to quality at every level may have a far greater impact than conspicuous expenditure on a few large-scale projects'[4].

Figure 5.3 shows some of the overall relationships of factors relating to the amenity of places.

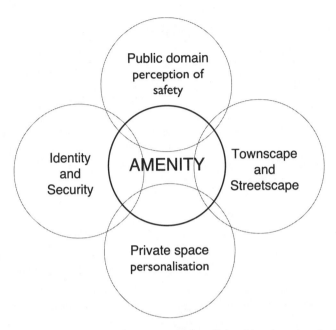

Figure 5.3 Amenity results from a variety of qualities (developed and adapted from *Vital and Viable Town Centres: Meeting the Challenge*, HMSO, London).

These three factors – attractions, accessibility and activity – are used by the DoE to develop an indicative health-check survey. The reader is recommended to refer to the report [4] for full details.

The DoE's framework of the 'three As' serves to illustrate very clearly the earlier point in this chapter that vitality is both a product of physical quality and variety and the activities that are generated by way of a response. The remainder of this section looks at the way some of the qualities which create vitality are changing, and not always for the better.

The twentieth century has experienced an unprecedented acceleration of the rate of change in the built environment and many people's lifestyles. These changes affect the character and qualities of towns and cities. Some of the factors influencing the variety and vitality of places are suggested below.

THE CHANGING QUALITIES OF TOWNS AND CITIES

WORKPIECE 5.1

INCREASING VITALITY

Design and undertake a survey of the pedestrian and cycle movement in your own or chosen town centre. You will need to identify:

- the location and variety of major land use generators including those near the edge of centre and clusters of specialist facilities;
- the location of access gateways such as public transport nodes;
- the use of park-and-ride or shuttle bus services where available;
- occupancy of off-street car parks;

- the frequency of use of public squares and parks in or accessible to the centre;
- the spread of activity over the whole day and by time of the week.

Design the creation of count points around major generators and access points.

Try to compare your results with the theoretical capacity of pedestrians and cycle ways and judge whether perceived activity is broadly based or, for example, purely retail.

QUALITY PUBLIC SPACE

The impact of a good public environment on image and attraction is enormous. Centenary and Victoria Squares in Birmingham illustrate this vividly. **Pedestrianization** in particular has been a central feature of our spatial thinking for some years. The evidence in the UK and Europe, albeit limited, does show that pedestrian flow and retail trade increase as a result [6]. However, there is a developing view that the complete elimination of traffic is not the answer in all circumstances. There are several reasons for this. Spaces themselves too often lack informality and are overplanned and detailed; economic decline from out-of-town development demands improved central access for at least some vehicles, which can then be restrictively managed; and vehicle movement is often seen as making streets safer.

Where attractions are strong, as in a historic town, or there is a vibrant evening economy, as in some bigger cities, the removal of traffic complements rather than deadens these attractions. But the everyday industrial or market town needs to avoid what URBED has defined as the pedestrianization trap. A different balance is needed, as in the new Carfax in Horsham, where space for pedestrians and vehicles is delineated simply by different but high-quality paving materials.

SHOPPING AND TOURISM

Shopping itself is becoming a leisure activity, both for the more affluent, who are discriminating and look for specialist independent shops and services, and for others. Places that exhibit vitality (and not just our historic towns) have usually created a rapport between this type of shopping and their other attractions in the fields of the arts, culture and entertainment, as in Nottingham and Bradford.

THE STAGE-SET

Off-peak animation of the built environment is self-evidently a function of mix and variety of use. The cultural losses in recent years in terms of cinemas, theatres and dance halls are now being offset by new attractions and experiences, such as hands-on museums, art centres, public art and sculpture, and the opportunity to witness industrial production like jewellery making (and buy the product). Ethnic restaurants are spreading, and there is every sign that pavements will be colonized for eating and drinking in the right conditions. The great and unused asset of our centres is the library, deplorably threatened by public spending limitations. The stage-set potential to use our streets and spaces for festivals, events and celebrations is also very great and as yet unrealized.

WORKPIECE 5.2

COMPARING ATTRACTIONS

Analyse and assess the comparative value of the range of attractions in your selected town centre.

Evaluate the contribution that each factor makes to the level of vitality (as measured in Workpiece 5.1).

The factors to consider and, where possible, to quantify will include:

- the scale, mix and quality of shopping and the degree to which retail sales are gained or lost by the centre;

- the presence of cultural, leisure and arts facilities which particularly attract visitors and tourists;
- the civic and educational role of the centre;
- the level of business and related services;
- the scale of residential uses.

Assess the degree to which each or all of these may need strengthening. See if you can generate some draft ideas as the basis of an action plan to achieve these changes.

PUBLIC TRANSPORT

Vitality is, in part, a product of our willingness to impose restrictive management of traffic in the interests both of people and of economic activity. The de-urbanization of the UK, fostered for years by the increasing spread of development, increasing car ownership and the consequent road building and decline in public transport, has created circumstances where traffic management alone is not enough. Belatedly, new directions are being set. De-urbanization will perhaps be tempered by Planning Policy Guidance Note 13 on Transport [7] and by emerging ideas of sustainability and compact city form. The trams are rolling again in Manchester and Sheffield, and park-and-ride is coming into its own in places like Oxford and Exeter, despite the millstone of bus deregulation. The integration of public transport into a total transport package and the attractive conditions it creates are fundamental to vitality, a relationship taken for granted in continental Europe. Why do they succeed where we and the Americans do not?

WORKPIECE 5.3

VARIATION CAUSED BY ACCESSIBILITY

Undertake a comparative desk study, based on a local authority survey, of the importance of accessibility to the vitality of town centres (including the centre used in Workpieces 5.1 and 5.2).

Use either a selected sub-region or towns of a particular size (50 000+) or type (e.g. larger market towns, former industrial towns).

For each, consider:

● number and frequency of bus and train services (using timetables);

● the penetration of these services into the centre (using maps to identify bus and train stations);

● the availability of minibus/shuttle/park-and-ride services;

● the level of pedestrianization (create an index of access: 0 for nil pedestrianization; 5 for maximum, i.e. all main shopping streets);

● the level of street parking, including multi-storey.

Compare your results with census or other survey data on the level of retail turnover. Look for the degree to which the scale of pedestrianization and public transport availability correlate directly or otherwise with vitality.

INTERNATIONAL EXPERIENCE

So far we have considered variety and vitality with reference to the UK. How does the UK experience relate to other parts of the western world? Some of the lessons from Europe and North America are described below. Do any of these lessons have relevance to other, less developed parts of the world?

THE EUROPEAN LESSON

The town and city centres of our European neighbours invariably feel much livelier than their British equivalents, despite the fact that many of them, particularly in the older industrial north (for example, Lille in France or Charleroi in Belgium) have faced similar problems of economic decline. This is not accidental, nor do they have a bourgeois tradition that Britain does not, but there are different pre-conditions. Densities are, in most cases, considerably higher with more people living close to the centre and, therefore, within walking distance. A high proportion of shops are independent and institutional control is much less. Apart from French hypermarkets, superstores and retail parks are rare and even a large city like Frankfurt has almost no equivalent to one of our major chains.

The system of regional and municipal government is also conducive to far more civic concerns about local property, with control of competition.

The policy differences are important as well. Most cities, and many towns, have invested in integrated public transport systems, including cycling, which are environmentally sound and well related to higher land use densities. Pedestrianization levels are higher, extending often into secondary streets which are then taken over by the cafés, restaurants and traders. There is a tradition of underground car-parking, rather than multi-storey structures which are as rare as inner relief roads.

To add to it all, far greater priority is given to the importance of the built heritage and to familiar townscapes, which seem under much less threat than they are here. The result is a liveliness and vitality that we envy. But communities in Britain are beginning to demand 'liveability' or urbanity from our policies, and since we still possess a very considerable heritage, we are not without the ingredients that count. Obstacles remain, and the next section identifies a few of these.

The other side of the coin is the other side of the Atlantic, especially in the United States, whose businesses and cultural ideas we seem to have imported so assiduously. Variety and vitality have fled from many American centres, where car-based dispersal has been relentless. 'Edge cities' and 'techno suburbs', some of them distinctive in their own right, have meant that retailing is no longer the prime downtown function [8]. Offices with some attractions for tourists who usually need to be told when they are entering the historic district, plus the inevitable car parks, have not turned the tide and the result is the phenomenon known as the doughnut.

New programmes such as the Main Street revitalization scheme in Canada are now supporting local initiatives to use the resources that remain and re-emphasizing the uniqueness of the central area, under a shared vision. Advantages are now seen in the direction followed by continental Europe because the undermining of the centre and the creation of a citadel undermines the financial basis of government and increases community dependence at just the time when global indicators would suggest the opposite is needed.

THE OTHER SIDE OF THE COIN: THE NORTH AMERICAN EXPERIENCE

The dispersal of land uses and the lowering of urban densities condoned by British planning has not reached the US level and options are still open – just. The health and vitality of UK centres has, until relatively recently, been badly neglected, whilst there has been an obsessive interest in rural and countryside affairs. This is partly a result of the marginalization of the planning process in the UK in the 1980s and is not simply the result of negative or passive planning.

In the 1990s, there has been an increase in planning influence in the UK which has come from two directions:

● the statutory change which linked presumption of approval of a

THE INFLUENCE OF PLANNING

planning application to a Development Plan allocation and ushered in what many now call the plan-led system;

● the revised policy guidance set out in the Government's planning policy guidance notes which emphasize the relationships between land use, transportation and the environmental and economic impacts of development.

At first sight, these directions may seem somewhat removed from variety and vitality. In fact, both are fundamental to achieving these qualities. The first has allowed planning to play a more positive, creative and enabling role – for example, promoting mixed uses and ensuring better quality. The second represents a quite fundamental shift in thinking towards urban concentration, and to town and city centres in particular, as preferred locations for retail and other investments.

Some UK guidance follows the emergence of clear evidence about the damaging impact of out-of-town retailing and the completely unsustainable direction it represented. Consideration of the impact of development proposals on vitality and viability is now mandatory where once (astonishingly) it was not seen as part of the planning process. Government has, in fact, tended to retain its encouragement of competition but now sees the future as one of planning for choice [9].

Other guidance is more comprehensive in its derivation from concern about air quality and the consequent aims of reducing the need to travel. It therefore brings land use and transport planning, including density, firmly together and provides for the first time a basis for the improvement of public transport. Similar problems and policy responses can be found in various places around the world.

WORKPIECE 5.4

IDENTIFYING LOCAL POLICY FRAMEWORKS

Make a checklist of the factors which influence variety and vitality in your locality, possibly by referring to local newspapers, and describe the prevailing policy framework and any steps which could be taken to promote vitality.

Considerable impediments remain despite the new directions set by policy frameworks. The potential influence of development plans, for example, may be less than it could be, if planning authorities lack the will or the resources to visualize these plans and give a design lead to developers. Relationships between authorities and with institutional town centre investors are sometimes meagre. Urban design as part of a local plan might not entirely overcome apparent institutional indifference to central decline and to the loss of the specialist traders who cannot afford the new rents at review time. But it would improve dialogue and understanding and, at least, improve the quality of what is actually built. Important heritage and townscape facilities that the local authority (and the community) regard as sacrosanct could be identified, and something could be done about the inflexible imposition of corporate image, which even the US is beginning to tackle (Example 5.1).

EXAMPLE 5.1

CORPORATE VISUAL RESPONSIBILITY

Corporate approach to marketing and design is too often visually unyielding and insensitive and leads to an erosion of distinctiveness. A new study in the US (*Saving Face: How Corporate Franchise Design can Respect Community Identity*, by Ronald Lee Flemming of the Townscape Institute [10]) has brought together a large range of examples (not confined to the United States) of design alternatives to standard prototypes. It demonstrates very effectively how community design review processes can require or persuade the corporate sector to modify in order to enhance and respect.

The report selects fast-food restaurants and filling stations as the most prominent corporate design influences in the US. Here, we might have looked at supermarkets and hotels or business parks. It acknowledges that any examples involve judgements about design appropriateness but makes the crucial distinction between good design *per se* and good design in context. Companies want an instantly recognizable image, but the report demonstrates how persistence and community-based design techniques can respect both a corporate and a community purpose. To take the ubiquitous McDonalds as an example, the Art Deco restoration in Figure 5.4 of a former clothing store in Halifax would be regarded in the UK as a good one, but hardly competes in quality with Walnut Street, Philadelphia (Figure 5.5) or Bolzano in Italy (Figure 5.6).

Figure 5.4   Halifax, England.   **Figure 5.5**   Philadelphia, USA.   **Figure 5.6**   Bolzano, Italy.

The techniques identified in the report for a community design input are unfamiliar in UK terms, but a possible model of what is needed to improve on our planning process. Five successful examples of community enhancement are analysed in the report. Taking one of these, Carmel in California, the seven community members of a planning commission, of whom at least one has to be a design professional, are appointed by the city council to serve four-year unpaid terms, meeting twice a month and reviewing an average of 14 cases per meeting. Carmel has strict sign ordinances but the commission's influence is still remarkable and the results tangible, without loss of variety or vitality.

SUSTAINING VITALITY

Variety in the built environment may be something we understand in the sense of recognizing it when we see it. The phenomenon of vitality, on the other hand, is more complex and elusive. Both are difficult to sustain, as they are affected not simply by physical wear and tear or neglect, but also by forces of social and economic change, often outside local control.

Many of the environments that we admire are the product of centuries of evolution, much of it informal, haphazard or accidental, and possessing qualities we now try to emulate by conscious planning. Variety, particularly on the small scale, seems to engender a human response and so leads to distinctiveness and vitality, which as a result can be created for even the biggest developments, like Broadgate in London. Difficulties and arguments develop where not only is there a striving for informality but also architectural styles are of the period. In Canterbury, for example, there is a central shopping redevelopment scheme that is pure pastiche, but successful and popular for all that.

EXAMPLE 5.2

BROADGATE/LIVERPOOL STREET

Broadgate is a speculative development with a difference. It demonstrates how a mixed use development can be planned to promote a vibrant and varied street life. Built in the 1980s on the site of Broadgate station and car park on the edge of the City of London, Broadgate offers office accommodation of the highest quality planned around three newly created traffic-free public squares.

The squares contain fountains, mature trees, sculpture and seating, offering an attractive and restful haven from the clamour of the surrounding streets. Bars, restaurants and cafés open on to these squares; an arcade of small shops has been incorporated; and a central amphitheatre forms an outdoor skating rink in winter and a stage for concerts, exhibitions and open-air theatre in

the summer. This careful combination of 'attractions' and 'amenities' means the squares and spaces between the buildings are always full of people and activity. 'Access' too has been brilliantly handled: Broadgate's spaces link directly to the refurbished Liverpool Street mainline and underground stations, while there are integrated circulation arrangements for buses, taxis and service vehicles.

The scheme has proved both a popular and commercial success, with 95% of the three million square feet of office space constructed since 1985 occupied. Broadgate is now 'home' to 25 000 workers. Its success is a damning condemnation of the many empty and dreary office plazas that have disfigured our town centres, contributing nothing to their health and vitality.

Arguments about whether an environment is better because it is largely unplanned are rather academic. Contemporary change and pressure have to be responded to, and it is important to:

- use the qualities and features already available in the built environment, which give continuity and stability, as a basis for new thinking and new urban design;
- develop a strategy;
- take action to restore vitality when it is weak;
- monitor the situation so that it is sustained.

Successful urban regeneration in Britain has been based on exactly this approach, as in Halifax (Example 5.3). In Birmingham a strategic and concentrated approach has changed the city's image by not only creating high-quality public spaces, but also by promoting distinct quarters of the city along Parisian lines, and downgrading the infamous Inner Ring Road to allow pedestrians to cross at grade.

To the basic principles outlined above, it is possible to add another: that it is necessary to recognize the importance of small-scale incremental change as a catalyst in achieving wider impact and recreating civic pride and community momentum. This is a means of both bringing vitality back and utilizing what already exists. There are countless examples of the dividends it brings, as with the simple but dramatic restoration of the footbridges over the Grand Union Canal at Maida Vale in London, or the famous restoration of Wigan Pier, both of which have prompted regeneration on a wider basis.

What else can be done to promote variety and vitality in the future? Two important new components have recently been added and these are described below. Could there be other helpful initiatives?

FUTURE DIRECTIONS

THE TOWN CENTRE MANAGER

The lack of coordination between agencies and departments of a local authority is a frustrating phenomenon of many centres and has much to do with poor maintenance and neglect. The response has been the simple idea of the town centre manager. There are now over 50 such posts, but most do more than simply coordinate and some are central to the creation of vision or appropriate action plans. In Nottingham, for example, town centre management has not only helped to bring cafés to pedestrianized streets; it has also helped to make bus-stops and car parks attractive and safe, giving the centre an almost continental feel.

EXAMPLE 5.3

HALIFAX: VITALITY FROM HERITAGE

The return of vitality to Halifax is an example of the way in which an area's history of manufacture and trade can be used to provide a new foundation for prosperity, and its 'unrivalled legacy of architectural and environmental riches', re-deployed to create visual variety and an upsurge in attraction.

The decline of the industries that once brought prosperity to Halifax (wool textiles, heavy engineering, carpet making) brought with it urban dereliction, unemployment and an outward loss of trade. The strategy that evolved had the following characteristics:

- the use of the environment as an asset in producing economic revival and a new image;
- a range of low-key schemes to give cumulative effect and recreate confidence;
- a partnership incorporating the community, the private sector, the DoE and the local authority.

The environmental approach involved a programme of shop-front restoration, the improvement of the Victorian market and the restoration of a small neglected central square. This simple process of good design improved appearance and increased retail sales, and was complemented by partial pedestrianization and improved street paving.

These important yet relatively inexpensive changes would have been impossible without two other decisions: first, to prevent an overscaled and potentially disastrous central area development that would have swept much heritage away; and second, to prevent the loss of the seventeenth century Piece Hall, a remarkable space of European quality whose redevelopment could scarcely be contemplated today.

The strategy was also helped by the entrepreneurial reincarnation of a massive former carpet mill at Dean Clough as a thriving industrial and cultural park. It is in such circumstances difficult to know what leads and what follows. Dean Clough would have had less impact without the town centre design programme, which in turn may have taken longer to create an encouraging investment climate without Dean Clough.

The important lesson is that variety and vitality have come from retaining and re-using the heritage of a place and new investment has followed, with an award-winning bus station incorporating three former Victorian façades into a remarkable modern design and, most recently, the building of the Eureka Children's Museum from private funds. Without the work of the 'Inheritance Project' (as it became known), however, such a symbol of a new future would never have come to Halifax. The lessons are clear.

THE URBAN VILLAGE

The characteristics of the 'villages' that once gave structure to our larger urban areas, particularly in London, are even today seen as offering lessons. The mixed uses, smaller scale and local facilities stand in stark contrast to the formless and unwelcoming nature evident in most larger housing developments of the last 30 years, whether public or private. The urban village is an initiative designed to bring variety and mix back into contemporary development, not just in greenfield situations, but within urban areas as well, where the concept is a tool for restructuring our cities in line with Planning Policy Guidance Note 13 and a basis for successful regeneration.

EXAMPLE 5.4

LIVING OVER THE SHOP

One of the striking differences between British and continental European towns and cities lies in the number and proportion of people actually living in the centre. It is no accident that those continental cities whose liveliness, vibrancy and variety we so much admire – Paris, Berlin or Amsterdam, for example – all have large and diverse residential populations in their downtown areas.

By contrast the proportion of people living in the central areas of British towns and cities is vanishingly small. For example, Leeds (which is by no means exceptional in this respect) has a total population of 700 000 of whom only 900 actually live in the city centre itself [11].

Yet, ironically, there is a massive amount of vacant accommodation in the central areas of our towns and cities, located above shops and other commercial premises. Some estimates put this as high as 500 000 potential dwellings. Set up in 1989, 'Living over the shop' (LOTS) is an initiative which has sought to bring this accommodation back into use.

One of the principal obstacles to bringing vacant upper floors back into residential use is the reluctance of most commercial owners and leaseholders to grant direct residential leases of any kind. Owners and their agents believe such leases devalue their properties and create potential management problems.

LOTS has pioneered a solution to this problem involving a two-stage leasing arrangement:

- The first stage is a commercial lease between the property owner (or leaseholder) and an intermediary, such as a housing association.
- The second stage is an assured shorthold tenancy offered by the housing association (or other intermediary) to the tenant.

Where housing associations act as intermediary, they have the advantage of being able to obtain public funds from the Housing Corporation for renovation, conversion and management.

This mechanism has proved successful in overcoming the reluctance of owners and leaseholders to allow residential use of their upper storeys:

- Commercial leases mean that the intermediary has full repairing obligations (including dilapidations) and is also responsible for all management matters (including rent collection).
- Assured shorthold tenancies mean that tenants can be removed and vacant possession obtained at the end of the lease – or earlier if there are problems.

By this means 3500 dwellings have been created, housing more than 7000 people. As well as numerous one-off schemes, several towns, such as Stamford in Lincolnshire and Ripon in North Yorkshire, have established rolling programmes.

LOTS schemes bring a number of benefits. The retailer gains rental income (after management and maintenance costs have been deducted) and also benefits in not having to pay unified business rates on the previously vacant upper storeys. The presence of tenants upstairs also means that the retailer's business premises are more secure and that the retailer is seen to be acting in a socially responsible way.

LOTS also brings wider benefits:

- Much-needed rented accommodation is provided.
- The existing building stock is repaired and kept in good order.
- The town centre becomes a livelier and safer place out of business hours.
- 24-hour occupation means that all neighbouring business premises are more secure.
- Revitalized town centres increase trade.
- There is a reduced need for building on greenfield sites.

SUMMARY

The urban village, with its concerns to recreate values that established variety and vitality in the past, is a positive point to conclude this chapter. It symbolizes welcome current concern with the need to use simple concepts to improve the built environment and so contribute to regeneration and to the creation and maintenance of delight and pleasure.

Variety and vitality are qualities at the heart of this regenerative process. The elements that give expression to them are now better understood and capable of being used in a particular place in a relatively systematic way.

So far, so good. Recognizing what is needed is vital, but not enough by itself. Vision, strategy, management and commitment to either renewing vitality and spirit or ensuring it is sustained are equally important. Resources are not necessarily the problem, bearing in mind that it is small things that make a difference, get things going and bring in resources behind them.

Above all else, the key to change is our new-found recognition of the importance of the health and well-being of our cities and towns, and the contribution that variety and vitality can make to their achievement. Let us hope that, soon, we shall no longer find it necessary to cross the channel to find them.

CHECKLIST

The issues considered in this chapter include:

- exploration of the factors which contribute to the variety and vitality of places;
- how these factors, and the forces that influence them, interrelate to encourage or reduce variety and vitality in different places;
- how changes in the development of towns and cities and life styles influence activity in different places;
- the lessons which can be learned from different places and cultures;
- some of the emerging initiatives which are being developed to encourage and enable places to retain or achieve variety and vitality;
- the importance of mixed uses in neighbourhoods and places.

REFERENCES

1. Girouard, M. (1990) *The English Town*, Yale University Press, London.
2. Department of the Environment (1992) *The Effects of Major Out-of-town Retail Development*, by BDP and the Oxford Institute of Retailing Management, HMSO, London.
3. Falk, N. (1995) *Successful Public Places: Going from Vision to Results*, Report, Vol. 4, May, pp. 16–18.

4. Department of the Environment (1994) *Vital and Viable Town Centres: Meeting the Challenge*, by URBED, HMSO, London.
5. Department of the Environment (1993) *Merry Hill Impact Study*, HMSO, London.
6. Hass-Klau, C., Nold, I., Böcker, G. and Crampton, G.L. (1992) *Civilized Streets: A Guide to Traffic Calming*, Environmental and Transport Planning, Brighton.
7. Department of the Environment (1988) *Highway Considerations in Development Control*, HMSO, London.
8. Garreau, J. (1992) *Edge City*, Doubleday, New York.
9. Department of the Environment (1988) *Major Retail Development*, HMSO, London.
10. Flemming, R.L. (1994) *Saving Face: How Corporate Franchise Design Can Respect Community Identity*, American Planning Association, Chicago.
11. Petherick, A. (1992) *Living Over the Shop: a guide for practitioners*, University of York, York.

Comedia (1991) *Out of Hours: Summary Report*, Comedia/Gulbenkian Foundation, London.
Hillman, J. (1989) *A New Look for London*, Royal Fine Art Commission, HMSO, London.
Worpole, K. (1992) *Towns for People*, Open University Press, Buckingham.

FURTHER READING

ENVIRONMENT AND SPACE

LES SPARKS AND DAVID CHAPMAN

THEME

How does the physical form of places contribute to the environmental and visual experiences of the people who use them? Are the townscape characteristics and the spatial qualities of the public realm important to our perception of places? Do they influence the environmental conditions of those places as well as the way in which they are used? This chapter explores some of the key factors and the way they may contribute to our understanding and enjoyment of places. It considers the appraisal and understanding of places and goes on to explore these in practice with reference to examples and case studies. Some of the themes explained in this chapter have been developed from Chapman and Larkham's introductory guide to urban design [1].

OBJECTIVES

After reading this chapter you should be able to:

● understand the possible effects of the physical form of places upon their users;

● recognize the interrelation of form, space and environmental conditions;

● consider how the same place may be differently perceived by different people at different times;

● develop techniques for area study and appraisal;

● consider the potential and value of 'urban design strategies'.

The public spaces in our towns and cities are an essential backdrop to our lives. We all have to pass through these areas on our way to school, to work, to the shops, and their form determines whether this will be an enjoyable experience or an ordeal. If it is an ordeal we may hurry through, eyes down, avoiding any diversion or human contact which might prolong the experience. Our thoughts and behaviour may tend to be anti-social. By contrast, attractive enjoyable spaces are more likely to encourage sociability, a willingness to pause for conversation or simply to greet an acquaintance.

What physical characteristics are likely to influence our attitudes to public spaces, to the streets and squares which form the corridors and rooms of our outdoor world?

At the most basic level, the degree of shelter and exposure to severe weather conditions is the first consideration. In northern latitudes, wide open spaces are usually cold and windswept and we often prefer the enclosure of surrounding buildings affording protection from wind and rain. Small squares and courtyards surrounded by buildings of two, three or four storeys provide good protection. Similarly, in hot climates, wide open areas expose citizens to the unrelenting heat of the sun. Again, closely arranged buildings and courtyards offer relieving shade.

The same principle applies to streets. Wide, straight and open streets have their place in most cities but, for sheer comfort, narrow streets with frequent changes of direction offer shade and protection from wind and rain. Add colonnades to the buildings or lines of trees in either a street or square, and further protection and comfort is provided. It is hardly surprising that our most popular towns and cities are often those whose public areas are characterized by networks of fine-grained streets, squares, courtyards and boulevards.

By contrast many 'modern' cities have replaced this close-knit fabric by high buildings standing in extensive open areas. The public spaces around high-rise buildings are notoriously uncomfortable, with the general lack of shelter being exacerbated by the turbulent wind conditions created at ground level by these tall structures. Such cities are often best experienced in the motor car, which itself insulates the passenger from the elements. In such circumstances the use and enjoyment of public space is minimal.

As a consequence of modern urban forms and motorized life styles, large areas of our towns and cities have become 'unsafe' – that is to say, citi-

INTRODUCTION

QUALITIES OF PLACE

SHELTER AND EXPOSURE

SECURITY

131

zens feel themselves to be at risk of violent assault, whether or not such assaults are prevalent.

The shape and form of outdoor spaces can contribute to a sense of relaxation or unease. Long narrow passageways, unless they are enclosed by occupied buildings, are oppressive and offer no chance of escape. Wider streets offer the option to cross to avoid strangers or to pass blind corners at a distance so as to avoid being ambushed.

Should streets and open spaces relate to a town's buildings and activities to ensure that they are well used? Busy places are usually safe ones, deserted places can be disconcerting, unless surrounded by occupied buildings. The stronger the relationship between a building's activities and the street, the safer the street becomes. Shops spilling out on to the street, residences with large window areas or balconies – these create a sense of security. Blank walls, empty office buildings, roller-shuttered shop fronts and narrow slit windows all convey a sense of anxiety and fear to the user of the urban space, if not to the building occupiers.

RESPONSIVE ENVIRONMENTS

A contribution to urban design theory was made in a now well-known design text, *Responsive Environments* [2]. This book suggests a range of principles for the design of developments, and the qualities to be aspired towards in every settlement. The key principles are legibility, variety, robustness, visual appropriateness, richness, personalization and permeability. Each of these principles implies increasing diversity and choice, and this is a key message. Is the quality of civilized life conditional on the variety of experiences and freedom available to all? As a manual for designers, the book contains many valuable insights into the qualities of 'good places', particularly the non-aesthetic qualities overlooked by most earlier theorists; but it also makes many cultural and ideological assumptions which may not be sustainable in the blanket universal fashion implied by the proposed formulae.

LEGIBILITY

One of the more alarming experiences is to become lost and confused in an unfamiliar city. It is important that urban areas have a recognizable structure and that different districts have identifiable features. Landmarks are essential aids to orientation and navigation in a large town or city – distinctive, tall landmarks are effective, particularly on the axis of a long street. But much smaller landmarks are also important: for example, unusual architectural features on the corner of street blocks where they can be spotted from a number of directions. The way people perceive places and find their way around was studied by Kevin Lynch [3] and some of his ideas are explained later in this chapter.

Views are the essential complement to landmarks. High vantage points are exploited best where they have a belvedere or open terrace from which to survey the surrounding area. Straight streets leading down across the contours from distinctive high points offer an advantage.

Street patterns in themselves can either aid understanding or create confusion. A simple grid-iron street plan, linked with a sufficiency of local landmark buildings, is usually an easy place to understand. Similarly, a settlement with a dominant central space from which the main streets radiate outwards is again an easy place to read. Most difficult and confusing are towns and cities which have neither of these characteristics, where streets intersect at odd angles, neither converging on important central places nor running in parallel, but laying down arbitrary and irregular networks. Ten minutes walking in such a city can leave one without any sense of direction. Yet many of our most popular historic towns, especially those of mediaeval origin, have just this composite character.

Subways and tunnels also create confusion. A short time spent underground can lead to a loss of bearings and, on emerging, one's new viewpoint is suddenly different from the last one. This can cause dramatic and memorable experiences where one emerges in a large square or on to a hillside with a striking view. But more often one emerges after two or three turns below ground on to an indistinguishable street.

Is it the subtle visual and spatial qualities which give places their appeal? Most texts have focused on these aspects of form and their historical antecedents. Devices of urbanism defined by Logie [4] are simple characteristics, commonly found in urban landscapes both by accident and through conscious decision. Many of these are spatial qualities, concerned with positioning of features and spatial interrelationships. Their significance, of course, lies in how we perceive them. Here are some examples:

VISUAL AND SPATIAL QUALITIES

- **Progressions** are a common urban landscape feature. Every street is a progression of sorts, but some are clearly defined with aesthetic reactions in mind. Straight streets bounded by buildings of uniform style, height and materials are common but, because these can be monotonous, elements of variety and surprise are important.
- **Surprise** is important, and much underrated. It can be experienced when formal streets open out into unexpected open spaces; when side-roads or entranceways reveal vistas or buildings, and sometimes more subtly with an unexpected sound or activity.

- **Contrast** is significant in urban landscapes which provide numerous variations in form, texture, spatial relationships, and even colour. Contrasts may be on the scale of an urban quarter, street or square, building or element.
- **Scale** is usually perceived in terms of the relationship between building (in whole or in part) and a human being. Elements sized for humans are referred to as being of **human scale**. The opposite, **monumental scale**, is used to create distinctive and important scale for key public buildings.

Camillo Sitte considered many of these concepts when he argued that city planning should be not merely a technical matter, but an artistic enterprise [5] . He claimed that 'anyone who has enjoyed the charms of an ancient city would hardly disagree with [the] idea of the strong influence of physical setting on the human soul'. Drawing upon precedent from Classical, Renaissance and Baroque periods, he proposed principles for the relationships between buildings, public spaces and the monuments which punctuate them.

UNDERSTANDING SPACE

Spatial qualities can be hard to understand or describe, but we all appreciate them. Some simplistic characteristics of space include the following.

- **Enclosure.** Spaces are usually most effective when they are clearly defined and provide a sense of enclosure between fairly well-defined edges. The level of enclosure and the proportion of the space all contribute to the nature of the experience and feelings of safety or pleasure.
- **Interpenetration.** Spaces are sometimes fully enclosed visually but not in form or plan. The linkages between spaces physically and visually can set up relationships between the spaces both horizontally and vertically (Figure 6.2 and Example 6.2).
- **Leakage of space.** Where a space is not visually enclosed, it is sometimes said to 'leak'. Usually, this is regarded as a weakness of the sense of enclosure, but sometimes glimpses out of a space to another space or the horizon can be a powerful experience.
- **Subdivision of space.** Elements within spaces may divide or subdivide the space, dramatically altering our perception and experience.
- **Abstract space.** Space as physically defined is readily perceived, but features and elements in the environment can also define 'invisible' space.

EXAMPLE 6.1

GENTLE STREET, FROME

Gentle Street (Figure 6.1) is a beautifully curved and steep pedestrian street paved with stone sets. The buildings which climb up the street at a steady pace, their windows looking out on to the public, closely hug the edge of the route to form a clearly defined channel.

Figure 6.1 Gentle Street, Frome (Les Sparks).

This is an example of how informal curves and slopes can add interest and hold out the possibility of surprise as the visitor proceeds along a path.

Individually, spaces may be attractive, but collectively a series of identical spaces could become monotonous. Ideally, an area should be made up of a series of positive contrasting spaces which are clearly defined and unambiguous. The shapes of spaces, and the proportion of their widths to the height of the buildings enclosing them, create very different experiences [6].

Focal points or nodes are 'destinations' where people may rest, meet or undertake activities. These are sometimes static spaces, in contrast with the dynamic spaces created by the linear streets and alleys which give the essential linkages within the whole structure.

CONTRASTING SPACES AND EXPERIENCES

135

EXAMPLE 6.2

INTERPENETRATION OF SPACES

Linkages between spaces both physically and functionally can set up interesting relationships. Figure 6.2 shows Newark Town Hall and Market Place. The Town Hall, built from 1773 to 1776, was designed with a covered market on the ground floor, extending the activities from the huge Square outside, through under the building itself.

Figure 6.2 Interpenetration of spaces: the Market Place, Newark (Les Sparks).

The articulation of urban form and space makes a fascinating study and is addressed by many commentators. The ways in which our appreciation of spaces are influenced are many, and include changes in direction and incidents along our way. The direction of streets is given by their shape or curvature, both horizontally and vertically (by topography and development), and their articulation by events or markers along the way.

SEQUENCE OF SPACES

The urban form of towns and cities is experienced through sequences of spaces as one moves around. Gordon Cullen called this 'serial vision' and described the variety of scenes one encounters moving through different parts of a town – through archways, along streets, past monuments, into squares, through narrow alleys, out on to open terraces, etc.[7].

Whilst the sequence of spaces in most traditional cities comprises a series of streets and squares, there is an infinite range of possible forms this can take, depending upon the local terrain, the activities in the surrounding buildings and their physical layout. Particularly in hot climates, the interior spaces and courtyards of buildings, although semi-private spaces, are at least as important as streets and public spaces.

A well-known illustration is the famous sequence of spaces created by John Wood in Bath (Figure 6.3). Here we see a steeply sloping street leading up from a rectangular space (Queen's Square) to a circular one (the Circus) on a plateau: one third of the way around the Circus, another street runs along level ground to the corner of a huge semi-elliptical range of buildings (the Royal Crescent) facing south across open parkland sloping away towards the river valley. This is a particularly sublime example, but an analysis of the variety of shapes and functions of urban squares demonstrates the scope for the urban designer.

SQUARES

The large public open space – square, place or plaza – is an important and much-studied aspect of urban form, function and design.

The two dominant influences on the character of town squares are shape and function. Function can vary from a private central garden, secured with railings for the sole enjoyment of the residents in the encircling houses, to open public spaces which may, in turn, become open-air markets, settings for ceremonial occasions (religious or secular), public gardens perhaps with fountains and sculpture, or paved civic piazzas with outdoor cafés and informal recreational activities (from boules to busking).

Shape can vary from large and monumental, to small and intimate, from irregular in plan form, to formal and geometric. Most squares are built on level ground, but many of the more memorable examples (e.g. in Tuscany) are created on sloping ground. Some squares provide the setting for important civic or religious buildings, either built along one side of the space and facing into it, or sometimes built within the space itself. English marketplaces have often been colonized by later churches, public buildings and shops – markedly changing the place's character and appearance, if not function.

Even the most common rectangular square can vary enormously according to whether the corners are closed, with streets entering at mid-points along the sides, or whether streets enter at the corners. One of the subtleties of Bath's Circus is its subdivision into thirds, so that on entry from any of the three streets one is confronted by a concave building façade (rather than looking straight through and beyond as would happen if it was divided into quarters).

Generally speaking, public squares form the truly memorable elements of successful towns and cities, but there are some remarkable streets too.

EXAMPLE 6.3

SEQUENCES OF SPACES: BATH

Queen's Square, built on a gentle slope, is linked by the more steeply sloping Gay Street to the Circus, built on a plateau. The view back down Gay Street affords one of Bath's famous sights. Brock Street continues along the contour without giving any indication of the splendour to come. On arriving at the end of the street, the empty horizon seen from the Circus turns out to be a vast theatrical arena. The Royal Crescent, built as a semi-ellipse, faces south with views across open parkland sloping down towards the valley. Marlborough Buildings to the west provides an effective windbreak against the prevailing south-westerlies.

The whole sequence is built in gold Bath Stone in the finest Georgian architecture with a subtle interplay of classical orders, providing a coherence and uniformity enriched with delightful detail (Figure 6.3).

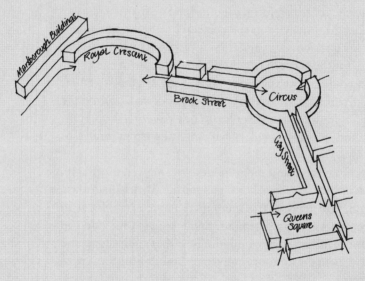

Figure 6.3 Sequences of spaces, Bath (Les Sparks).

STREETS

Streets, like squares, are influenced by their shape and function. Function results both from the type and volume of traffic they carry, and from the activities in buildings along their length. The volume of traffic, whether pedestrian or vehicular, determines the character and attractiveness of a street more than any other factor.

The form of streets can be as varied as that of squares. The cross-section through the street (i.e. the relationship between width of street and height of adjoining buildings) is a critical factor. So, too, is alignment: straight streets are sometimes associated with monumentality and civic order, whereas curving, twisting streets are likely to be more attractive and picturesque – a more enjoyable visual experience for the pedestrian as sequences of views and incidents are revealed.

The relationship between streets and buildings is just as important as that between squares and buildings. Streets benefit from an interesting building closing off the end view and from intriguing events along the way – towers, arched glimpses into courtyards, side-streets, etc.

Architecture and the urban form are experienced through all of our senses. The conscious and unconscious reactions to places and buildings are complex and quite individual, but various authors have explored the main factors which influence our reactions. In *Experiencing Architecture*, Rasmussen [8] offered an excellent and accessible introduction to some of the factors at work. He stated: 'The architect works with form and mass as the sculptor does, like the painter with colour. But alone of these [it] is a functional act. It creates tools for human beings and utility plays a decisive role in judging it.' Nevertheless, it is mainly the visual and sensual qualities which Rasmussen explores. He also reminded us of the effect of our experiences upon our perception of objects, their hardness or softness, their rigidity or flexibility, their weight or lightness, and the way in which these experiences influence our appreciation. The following characteristics are explored by Rasmussen.

PLACES AND BUILDINGS

- **Form and mass.** Like sculpture, buildings have both form and mass. The building up of the differently shaped parts of a building are described as massing, and form the edges of urban spaces (Figure 6.4).
- **Solids and cavities.** Buildings may possess physical form and mass, but they also enclose space within them and delineate space without. There is a positive–negative relationship between solids and cavities.
- **Proportion and scale.** Proportion is simply the relationship or ratio between the size of the elements of objects, but this has a significant effect upon visual appearance. Various attempts to discover an ideal system of proportion have been made, notably the Golden Section and Le Corbusier's *Le Modulor* [9].

EXAMPLE 6.4

PLACES AND BUILDINGS

The architect works with form and mass as the sculptor does ...[7]

Like sculpture, buildings have both form and mass. Boston City Hall demonstrates the architectural qualities of massing, solids and cavities, proportion and scale, and rhythm. It is the dominant structure overlooking Boston City Hall Plaza with its cascade and pool. The scale of the building (particularly the vast size of the lower storey elements) affords a monumentality befitting a civic place.

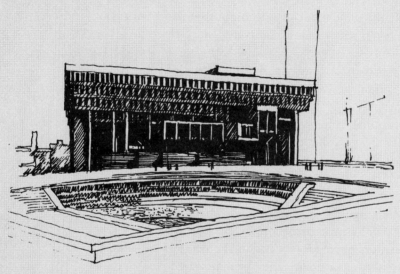

Figure 6.4 Places and buildings, Boston City Hall (Les Sparks).

- **Rhythm.** Musical metaphors are common in architectural theory. Rhythm is the frequency of repetition of the elements which articulate architecture and space. They are created by structures/voids and windows/walls as well as the subdivisions of building surfaces and the positions of street furniture.

- **Textures and materials.** Each material possesses its own qualities and characteristics, including texture. Combinations of different materials in buildings and surfaces create different effects. Are traditional small-scale elements related to the human scale particularly pleasing in visual or psychological terms?

- **Light.** The path of the sun across the sky is predictable. The orientation of streets and the design of buildings can respond to this and create places of both environmental and aesthetic quality. In warm climates, shade will be welcome; while in cooler latitudes, a sunny aspect will be desired. The unpredictability of atmospheric conditions and the changing effects of natural light through the seasons, and throughout the day, give us rich and varied experiences.

- **Colour.** Colour can be used to modify or highlight both form and space. Perhaps as important is the effect of materials and their natural coloration upon the character of distinctive areas. The inappropriate use of colour in an environment can be offensive, but the inspired use of a splash of colour can be very exciting.
- **Hearing.** We sometimes hear what is going on within buildings – for example with industrial processes, concerts and worship – but a powerful effect upon our hearing is caused by the shape of buildings. The volume of space and the hardness of surfaces together reflect sound and affect its reverberation.

Is it the variety of these factors, their contrast and combination, which make them so pleasurable a part of our experience of the urban environment?

WORKPIECE 6.1

APPRECIATING ARCHITECTURE

Select a building which you really enjoy seeing or being in.

Try to identify and describe the characteristics which cause your positive response to the building. You may use some of Rasmussen's ideas to help you [8].

Try the same excercise with a building you do not like or enjoy.

The exercise can be tried again and the factors which could arise are almost limitless. Consider spatial and environmental conditions, design creativity, technological innovation and contextural relationships, as well as the form, mass and visual qualities of the building.

THE ANALYSIS OF PLACE

Students of the built environment professions will all be introduced to techniques of site analysis. Careful survey, analysis and appraisal are essential to site planning, financial evaluation and design. While most disciplines are challenged with a specific site and proposed use, the planner and urban designer are usually confronted with huge areas, containing multiple sites and buildings, with a diversity of potential uses and activities.

Techniques for area appraisal can be used in a variety of ways. Each situation will deserve a unique approach to reflect the existing circumstances and the reason for intervention. The challenge to conserve a neglected historic townscape requires a different approach from that needed for the regeneration of a derelict and abandoned industrial area. The qualities of place ultimately desired may be similar.

141

Kevin Lynch, in his book *The Image of the City* [3], describes a process for analysing the structure of Boston, Jersey City and Los Angeles. The technique involved drawing out the mental maps of the citizens of those cities through interviews, supplemented by photographic recognition tests, and by requests for directions from people in the street. A systematic field reconnaissance of each area was made by a trained observer who mapped the key structural elements of the place and their interrelationships, and made subjective judgments about the strength of these elements in the overall image of the urban area. This would be followed by lengthy interviews with a sample of residents to establish their own images of their surroundings. The interviews called for the performance of imaginary trips, describing the main physical features which marked the journey. The exercise demonstrated that substantial shared images can be established through this process, and that these closely matched the images constructed initially by the trained observers.

Lynch then analysed the visual form of the cities by mapping those features which were identified repeatedly through this process. He classified them by five terms: paths, nodes, edges, districts and landmarks; and accorded major or minor status to each element depending on its prominence in the minds of the residents and the visiting observers.

- **Paths** are routes which people may use and could be streets, pathways, canals, railways or highways. People experience their environment by moving through them.
- **Nodes** are destinations: places of activity and focus. It is suggested that nodes are intensive foci to and from which we travel; they may be squares, or road and path junctions, and are sometimes the foci of districts.
- **Landmarks** are seen as point references, of importance in recognition, familiarity and wayfinding; but, unlike nodes, it is not usually important to be able to enter them.
- **Districts** are described as broad areas which are recognizable as having some common, identifying character, into which one can mentally enter.
- **Edges** define spatial enclosure. They are linear elements forming boundaries between places, linear breaks in continuity, edges of development: they may be barriers closing one region from another, or they may be seams along which regions are related.

This work produces a very different sort of map or city plan from the conventional street map – but one which has more immediate use to the urban designer.

While this map identifies the strong points in the urban image, the research was also able to identify inherent weaknesses in the urban form which lead to confusion and lack of understanding about other parts of the city and how they fit together. A complementary map was drawn annotating these problem areas as 'confusions, floating points, weak boundaries, isolations, breaks in continuity, ambiguities, branchings, lack of character or differentiations'.

The analysis carried out in Lynch's mental maps and definitions of image problems provides a basis for creative decisions and interventions by an urban designer. Yet it is difficult to use these 'maps' – for they are not predictive, reflecting instead past experience [10]. They may have some utility in explaining processes of familiarity, wayfinding and landmarks, but the absence of a feature in a representation of a 'mental map' does not mean that the person is ignorant of it. They should thus be used with caution in the design process, though designers may continuously seek to create readily grasped urban forms.

DISTRICTS AND THEIR BOUNDARIES

Towns and cities usually contain a collection of distinctive districts. These districts are distinguishable by their topography or their historical period of development, by their predominant functions or predominant social groups, their building types and building materials, or by combinations of these. The distinctive characteristics of these different districts is something which helps to give structure to the city as a whole.

The extent to which a district is homogeneous and sharply differentiated from other parts of the city will be variable, as will be any certainty about its boundaries. Many districts merge into one another such that no definitive boundary can be identified, either on the ground or in the perceptions of the inhabitants.

In other cases a district may have a clear edge to it. Transportation routes – railways, major highways, rivers or canals – often form clear boundaries to districts. The more continuous they are, and the more impenetrable they are, the more they structure the city and provide hard edges to different regions or districts.

TOWNSCAPE

The term 'townscape' is often regarded as descriptive of urban form, but the principles have much more potential as an aid to analysis and design. Cullen claimed: 'There is an art of relationship just as there is an art of architecture. Its purpose is to take all the elements that go to create the environment; buildings, trees, nature, water, traffic, advertisements and so on and to weave them together in such a way that drama is released' [7]. He identified three components of the way we appreciate townscapes: serial vision, sense of place and content.

SERIAL VISION   This term means that the enclosure of urban space, with streets, squares and some more open places is 'revealed' to the user not as a continuum but in a series of experiences. The experience of a long straight street is similar from beginning to end, but arriving in a square or passing through an archway into a small courtyard offers varied experiences, in what Cullen described as a series of 'jerks' [7].

SENSE OF PLACE   'The whole city is a plastic experience, a journey through pressures and vacuums, a sequence of exposures and enclosures, of constraint and relief' [7].

There are emotional reactions to exposure and enclosure: being inside, entering or leaving spaces, being above or below. Places can be occupied or possessed by their users. They provide a wide variety of spatial and visual experiences and give identity to the area.

Other writers, including the urban morphologist M.R.G. Conzen, have used the term 'sense of place' or 'genius loci' to refer to the emotional attachment produced by familiarity with the physical aspects of a place and a knowledge of its history and continuity, thus providing a sense of location in both space and time [11].

CONTENT   The fabric of towns contains colour, texture, scale, style, character, personality and uniqueness. Older towns in particular contain a variety of styles, scales and materials, and display interesting juxtapositions of layouts (often regarded as 'accidents').

> The popular conception of planning is to create order, symmetry, balance, perfection and conformity. Conformity is difficult for the designer to avoid, and creating artificial variations can be worse than the original boredom, but within an accepted framework it is possible to manipulate the nuances of scale and style, texture and colour, character and individuality. [7]

THE FUNCTIONAL TRADITION   Cullen's concept of townscape extended beyond serial vision, place and content, and included the recognition of an important 'functional tradition' which he described

144

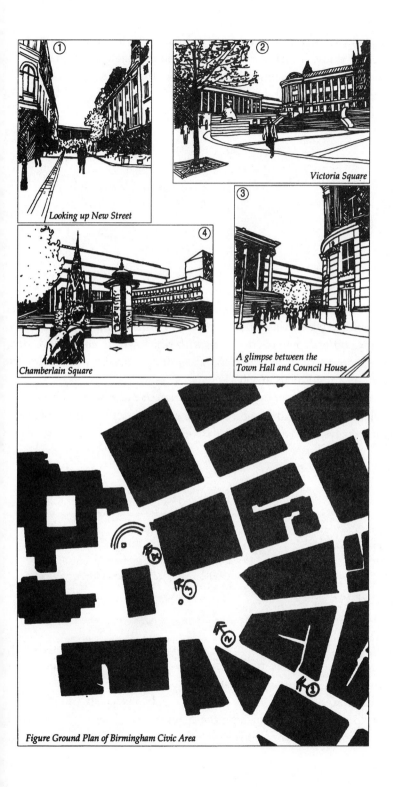

① Looking up New Street

② Victoria Square

③ A glimpse between the Town Hall and Council House

④ Chamberlain Square

Figure Ground Plan of Birmingham Civic Area

Figure 6.5 A serial vision sequence, Birmingham (David Chapman).

145

as 'the intrinsic quality of things made'. In essence, he means the simple elegance of things which are designed to do their job functionally and effectively, and which communicate visually the way they work or the message they are intended to convey.

WORKPIECE 6.2

STUDYING URBAN SPACE

There are many ways in which the form and character of urban space can be studied and communicated graphically. Pick an area with interesting 'townscape' and produce:

● a figure-ground plan of the area;
● a serial vision sequence [7].

Examples of both are shown in Figure 6.5.

ANALYTICAL GRAPHICS AND NOTATIONS

In *Notation*, Cullen [12] suggested an approach to townscape analysis which could be used by anyone interested in urban form. In practice, his ideas only address some of the issues with which we are concerned. It is to be expected that each study, appraisal, plan or design will inspire its own unique graphical approach.

Existing studies and plans demonstrate a variety of techniques. They all have four broad purposes, which relate directly to the processes of survey, analysis and appraisal.

WORKPIECE 6.3

ANALYTICAL GRAPHICS

Plans can be used as the base for many different area-based studies. Select an area for study and obtain a large-scale plan (say 1:500–1:1250). Develop your own graphic symbols and key for one of the following:

● age, style or 'value' of buildings;

● land uses and development/townscape 'opportunities';
● Lynch's five key elements of urban form [3];
● centres of activity and pedestrian linkages.

ANALYSIS OF URBAN FORM

Having identified some of the key factors which go towards defining urban form, it is possible to look at some of the techniques adopted in undertaking an urban design analysis of a town or city. Any visitor to a strange town tends to carry out this process subconsciously and in a rudimentary manner – distinguishing the main streets, locating the central squares, familiarizing themselves with the characteristics and location of

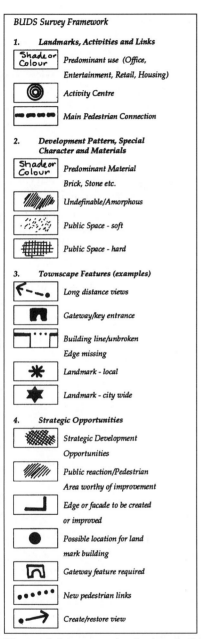

Figure 6.6 An example of urban design notation, developed by David Chapman from the Birmingham Urban Design Study.

major landmarks, and getting to know the broad divisions of the place into different districts with their respective functions and their ethnic or class structure. The mental maps which visitors formulate from their

147

early observations gradually become enriched with detail and enlarged as they stay longer and come to know a place better.

Residents, too, have mental maps of their home town or city. Residents from different districts will develop different mental maps according to their place of residence and the extent to which they use other parts of the city. For those key central parts of the city to which most people have recourse, there will be a common appreciation of the key structural elements of the place.

AREA APPRAISAL

One of the ways of improving our understanding of places and developing strategies for guiding development in them is the carrying out of an appraisal of quite wide areas, in order to help shape strategic objectives and locally sensitive policies for future action and development. Clearly, such studies must fit in with, and inform, the planning policies for an area, and will be concerned more with the physical and qualitative achievement of those policies. Lynch and Hack [12] suggested some useful objectives of analysis:

● seeking coherent patterns;
● understanding the place's equilibrium;
● discovering unique character;
● developing concepts based upon understanding of the place's distinctive characteristics.

Often the designer will be working with the grain of the locality, emphasizing its strong points, teasing out its potentialities. Sometimes [they] will dramatically cut across it or oppose its nature. [13]

SURVEYS

Surveys of an area can discover factual information and develop personal impressions. Research elsewhere generates information not obvious on the ground. It is important to identify key issues quickly, to focus survey and research work[1]. The following notes give an initial starting point.

● **History and background.** Examination of the history of an area and its archaeology, origins, structure and traditions provides an essential perspective on future opportunities for action.
● **Structure and relationships.** Geographical, geological, meteorological and hydrological structures need to be appreciated in order to understand how the pattern of movement, use and infrastructure have developed to take advantage of the physical environment and microclimate.

EXAMPLE 6.5

THE BIRMINGHAM URBAN DESIGN STUDY

In 1989, Birmingham City Council commissioned an urban design study following a symposium of international and local advisers, who considered how the city could best address the evident shortcomings of the city's core, making it a more 'user-friendly' place yet retaining accessibility and the potential to increase prosperity. This study clearly and simply set out to 'present a robust, coherent, apolitical vision of how the physical environment of Birmingham's central area can be gradually improved over the next 30 years or so, as opportunities for change occur' [14].

It was carried out by consultants, and the team included architects, planners, urban designers and a graphic designer. It is valuable as a framework or setting for future decisions, but it was not able to involve local communities and interests in the way which would be desirable in detailed decision making.

The study was organized under seven main headings:

HELPING PEOPLE FIND THEIR WAY AROUND It is interesting that this was the first section of the study, dealing with impressions of the city gained on arrival by various means of transport, and while moving around in the city centre. The permeability of the main railway station and the legibility of its signing, for example, was heavily criticized; and ideas were presented for the eventual Midland Metro links and for the removal of many barriers to free pedestrian movement.

DEVELOPING AND PROTECTING THE VIEWS This section identified four key landmark buildings which act as markers, and suggested that future tall buildings should adopt distinctive forms. Existing and potential views were listed, with suggestions for new physical forms to improve the potential views. The majority of these suggestions were for new façades or landmark buildings.

REINFORCING THE CITY'S TOPOGRAPHY This made the key point that the city is on a succession of ridges and valleys, suggesting strongly that this topography should be emphasized in the location and design of new tall buildings; and that all building heights should pay regard to contours.

REDEFINING THE STREET This makes the point that many of the post-war core redevelopments have introduced wide unused spaces between buildings and roads, removing traditional street forms and destroying human scale. Numerous buildings have no interest or vitality at ground level. Redevelopment should address these failings.

SWEEPING AWAY THE CLUTTER Here it was suggested that insensitively designed buildings should be re-fronted or removed. Advertising should be strictly controlled; shop front design should be more sensitive to the shop buildings; and street furniture and paving should be consistently of high quality. The removal of redundant and useless visual clutter was the keynote here.

SOFTENING/ENHANCING OPEN SPACE Planting is emphasized for areas in the city core, as few are 'green' owing to the historical development pattern. The canals are identified as a key network of open space to be enhanced, and the edges of these spaces should be marked by appropriate buildings.

REINFORCING THE AREAS OF CHARACTER All of these concepts are pulled together in an examination of 15 identifiable districts or 'quarters'. A plan and description for each contains suggestions for development which would reinforce the existing character of the quarters.

● **Human activities and behaviour.** There are often perceptible relationships between spatial form and behaviour, as human use responds to 'desire lines' and to the intangible qualities of atmosphere, emotional reaction and spiritual significance. Behaviour settings are those places which, because of their particular characteristics of convenience or quality, encourage a particular type of behaviour.

- **Qualities of place.** Places can possess both good and bad qualities. Understanding the way in which people perceive and react to places is an important ability for all those interested in the quality of places. Characteristics which will encourage human use and enjoyment include a range of attributes, from safety and security to visual and sensual pleasure.

- **Building form and space.** The importance of building form and the influence of spatial qualities upon our reaction to the quality of place and feeling of safety is debatable, but there does seem to be some connection. An appraisal of the spatial qualities of a place and the form and nature of groups of buildings is important.

- **Sensory qualities.** All of our senses and mental faculties are employed in interpreting our surroundings. Thus we assess the level of threat, the conditions around us and the sensory pleasure or offence being experienced. Noise and atmospheric pollution, exhilarating views and vistas, eyesores and smells all contribute to the sensory perception of a place.

None of these factors is static. Even the history of a place continues to 'evolve': the characteristics, qualities and activities of a place are changing daily, even hourly. Understanding these changes, their ebb and flow or their progressive mutation, is useful for anyone seeking to understand or guide the progress of urban change.

WORKPIECE 6.4

AREA STUDY

Select a study area which extends over several street blocks. It could be part of a town or a small village.

Using the techniques developed in previous workpieces carry out a study to identify:

- distinctive areas of character (and illustrate their key features);

- centres of activity and the principal links between them;
- the spatial and functional qualities of the principal streets and spaces forming those links;
- opportunities for development which might improve the urban form and 'townscape';
- opportunities to improve the 'public domain'.

When proposals are being considered to develop or otherwise change an area or site, it is important to carry out a detailed appraisal of the immediate area concerned, its context and all the factors likely to affect or be affected by the proposal. If a broad area appraisal has been carried out in an urban design study, this will provide a valuable framework for appraisal of individual sites.

For our purposes, a 'site' is the area defined as being an area detailed on a plan as the location and area for the 'change'. It is important to remember that the site only exists in relation to the proposal under consideration, and to remember that a multitude of other 'sites' could be defined in the same location in relation to other proposals. For example, a field being considered for development as a small residential estate could form part of a much larger site being considered for a major new out-of-town shopping centre.

It is also important to remember that not all of the site will necessarily be subject to change, and that other areas outside the site itself may need to be changed to facilitate the proposed development. There are two sides to every site boundary – inside and outside – and knowledge of conditions on both sides is necessary.

Site appraisal is equally valuable for developing strategies of managing sensitive ecologically important sites, as well as proposals for new landscape or building schemes. It involves much more than a factual survey. We need to ask what opportunities and constraints are presented by any given areas, and how these can best be exploited or overcome to achieve the desired results.

It is generally suggested that there are three main stages to site analysis:

- **Survey**: gathering information by observation, enquiry and research, from the site itself, from documentary records and from local knowledge.
- **Analysis**: consideration of the information acquired, examination of interrelationships, identification of constraints and opportunities. Often alternative strategies for achieving the desired outcomes will be generated, each with advantages and disadvantages.
- **Appraisal**: drawing conclusions about the site design and layout in relation to the desired objective. This will define clear aims and objectives, and possibly performance criteria by which success may be measured. It should also portray a vision for the future of the site.

These stages are not independent or sequential. They are part of an integrated and interactive process. The analysis may suggest the need for more survey data; the appraisal may prompt further research and analysis.

WORKPIECE 6.5

SITE ANALYSIS

The process of area or site appraisal can be very personal, and there is no correct method . Select a small site for a study, indicating graphically, on scaled plans, the results of the survey, constraints and opportunities and, where appropriate, actual proposals. All of the factors outlined for area appraisal will have relevance to any individual site appraisal. The following checklist provides a starting-point for any survey, but key issues should quickly be identified: remember that this list is not exhaustive, nor is every element equally important to all sites [14].

● Record general impressions. Use notes (perhaps taped), sketches, plans and photographs.
● Record the physical characteristics: site dimensions/area, boundaries, slopes, ground conditions, drainage, trees and vegetation, buildings and other features.
● Examine relationships between site and surroundings: land uses, roads and footpaths, transport nodes, stations, bus stops, local facilities and services.

● Consider environmental factors affecting the site: orientation, sunlight/daylight, climate, microclimate, prevailing winds, shade/shelter, exposure, pollution, noise, fumes, smells.
● Assess visual and spatial characteristics: views or vistas, attractive features or buildings, eyesores, quality of townscapes and surrounding spaces.
● Note any danger signals: subsidence, landslips, poorly drained or marshy ground, fly-tipping, vandalism, incompatible activities or adjacent uses.
● Observe human behaviour: desire lines, behaviour settings, atmosphere uses.
● Consider the area's background and history: local and regional materials, traditions, styles, details. Historical research will sometimes be required to explore the development of the area.
● Research statutory and legal constraints. Ownership, rights of way, planning area status, planning conditions, covenants.

Using these approaches to area and site appraisal, it is possible to develop strategic objectives and policies for a wide study area and for each distinctive area within it. Such strategies will both influence and respect the political, economic and planning framework, and they require clarity about the purposes of the study. Each study requires development of a unique approach, although each will consider the existing situation, desirable characteristics and possible actions or steps.

SUMMARY

We have approached this chapter with the intention of drawing out those issues and qualities which make places enjoyable. Inevitably, our origins in western Europe precondition both our values and aspirations. Nevertheless, some of the ideas for local appraisal have wide application.

The principles and ideas we have discussed are applicable to cities and villages world-wide.

Each country or region has its distinctive culture, but many are fast becoming multicultural societies. The vitality and variety of different cultures adds greatly to each society, both in the exchange of ideas and beliefs and also in stimulating new inspiration for art and design. This raises the important and unanswerable question of how fast we should change and embrace new innovations in urban form, and how much we should seek to preserve or conserve the forms of the past. This question can only be resolved at the local level, based upon a shared vision for the future of that distinctive place. We can neither afford to forget the lessons of the past nor shy away from the challenges and opportunities of the future.

The design process is fundamental to the achievement of our corporate and individual aspirations. It can enable the effective use and conservation of scarce resources, as well as the efficient use of land. It should harness technological advances and integrate new developments satisfactorily into existing settings. Design enables the objectives of the brief to be met and conflicting demands and requirements to be resolved. We argue that subtle urban design objectives and frameworks will enable even the worst designer to contribute to the overall form and quality of the place. Nevertheless, we should always promote and support patronage for the best possible design thinking.

This chapter has explored:

CHECKLIST

- some of the effects of the physical form of places upon their users;
- how form, space and environmental conditions relate to the quality of places;
- how the same place may be differently perceived by different people at different times;
- techniques for area study and site appraisal;
- ways of developing urban design studies and strategies;
- the value of both understanding and creativity in building neighbourhoods and places.

REFERENCES

1. Chapman, D. and Larkham, P.J. (1994) *Understanding Urban Design*, UCE, Birmingham.
2. Bentley, I. *et al.* (1991) *Responsive Environments: a manual for designers*, Architectural Press, London.

3. Lynch, K. (1969) *The Image of the City*, MIT Press, Cambridge, Massachusetts.
4. Logie, G. (1954) *The Urban Scene*, Faber, London.
5. Sitte, C. (1889) City planning according to artistic principles, translated by Collins, G.R. and Collins, C.C. (1986) *Camillo Sitte*, Rizzoli, New York.
6. Krier, R. (1979) *Urban Space*, Academy Editions, London.
7. Cullen, G. (1961) *Townscape*, Architectural Press, London.
8. Rasmussen, S.E (1959) *Experiencing Architecture*, Chapman & Hall, London.
9. Le Corbusier (1954) *The Modulor: a harmonious measure to the human scale universally applicable to architecture and mechanics* (translation), Faber, London.
10. Gould, P. and White, R. (1992) *Mental Maps*, 2nd edn, Routledge, London.
11. Conzen, M.R.G. (1969) *Alnwick, Northumberland: a study in town-plan analysis*, 2nd edn, Publication No. 27, Institute of British Geographers, London.
12. Cullen, G. (1968) *Notation*, Alcan Industries Ltd, De Montford Press, Leicester and London.
13. Lynch, K. and Hack, G. (1984) *Site Planning*, 3rd edn, MIT Press, Cambridge, Massachusetts.
14. Tibbalds Colborne (1990) *Birmingham Urban Design Study*, City Council, Birmingham.

FURTHER READING

Cowan, R. (1995) *The Cities Design Forgot*, Urban Initiatives, London.

Gosling, D. and Maitland, B.S. (1984) *Concepts of Urban Design*, Academy Editions, London.

Tugnutt, A. and Robertson, M. (1987) *Making Townscape: a contextual approach to building in a urban setting*, Mitchell, London.

Von Meiss, P. (1990) *Elements of Architecture: from Form to Place*, Van Nostrand Reinhold, London.

Worskett, R. (1969) *The Character of Towns*, Architectural Press, London.

154

CREATING PEOPLE-FRIENDLY PLACES

PUBLIC POLICY AND PLANNING

KEVIN MURRAY AND DAVID CHAPMAN

Civilization is, at least in part, a process by which the interests of all the people and life forms on earth are protected against the interests of any one individual group of people or species. This process is not easy. How can individual liberties be protected if they threaten the liberty or life of others? How can the natural desire to progress and achieve be balanced with the equitable distribution and prudent use of resources?

These high ideals are not necessarily the main influences upon the formulation of public policy and enactment of law. Vested interests and pressure groups variously influence law, policy and implementation. How do these forces and resultant frameworks of public policy affect people and places locally? How can individual and neighbourhood interests be represented in policy formulation? Public policy and planning are increasingly important in our densely and rapidly changing built environment. How well do they balance competing and conflicting interests?.

These are the sorts of questions which are explored in this chapter.

How can public policies and planning systems influence the quality and robustness of places? Do they encourage and facilitate achievement and involvement, or do they sometimes create barriers and obstacles? This chapter explains some of the issues raised by different policies and systems, and the different ways they can enable participation in decision-making processes [1].

OBJECTIVES

After reading this chapter you should be able to:

- understand the general influences on public policy as it affects planning the built environment;

- identify the key instruments of public policy in planning;

- identify the key players and their motivations;

- appreciate the strengths and weaknesses of conventional public involvement;

- examine an alternative contemporary approach to public involvement.

INTRODUCTION

Societies have increasingly adopted constitutions and laws for their governance. The instruments of government often provide national frameworks or local guidance for urban development and management. As land use planning systems have been incorporated into levels of central, regional or local government, so the public in whose interest the systems are devised come into contact with them. The interface with the public influences both policy and other decision making; it also influences the way in which the public – or different publics – can play a part in the system, whether formally or otherwise. This chapter explores some of the motivations underlying the foundation of policies and plans, the different levels of participation which might be found and contemporary techniques for fostering democratic involvement.

The experience and lessons in this chapter are drawn primarily from the British planning system [2]. However, many of the issues are relevant to other systems, particularly those of advanced industrial democracies. For other systems, where the regulation of land use is less open to public scrutiny and involvement, there are also some useful pointers for the future [3].

WHERE DOES PUBLIC POLICY COME FROM?

The policy which comes to regulate the use of development land in the public interest comes from a variety of sources and is expressed in a variety of ways [4]. The forms of policy may vary from place to place. For example:

- a part of the national legislative framework such as an Act of Parliament or Statutory Instrument;
- other forms of supplementary policy or guidance;
- government advice, for instance on 'good practice';
- statutory plans, policies and strategy documents adopted by local government;
- non-statutory guidance and policy, for instance on design issues or a specific site.

WORKPIECE 7.I

CONTEXTS FOR POLICIES AND PLACES

> Draw up a chart showing the origins of policies and plans in your country. Mark them with the levels and organizations which are responsible for them statutorily and show any activities which are non-statutory or advisory.

Most general procedures and policy appear to emanate from 'the top' (the State or the European Union, for instance) and 'trickle down' through the hierarchy of public policy making. In reality much public policy may be developed in a more complex way [5].

Theoretically public policy emanates from governments. Democratic governments secure their position of authority by the votes cast on their proposed programme or manifesto. Rarely does the regulation of land feature as high as other determining factors such as tax, employment, housing, health or education. Therefore, public policy on the regulation and land is not normally a feature of the debate about the future government of a country, except perhaps where radical reform (such as land nationalization) may be proposed.

Public policy on land use planning is therefore frequently open to a range of influences, including:

- the ideology and general political and social culture of the governing party;
- the attitude and interests of the responsible politicians (for these change even in the same political group);
- the key influences coming to bear on politicians (such as public servants, interest/lobby groups);
- 'bottom up' concerns by local authorities who administer the system;
- the culture, attitude and concerns expressed by the wider population in a range of ways.

WORKPIECE 7.2

THE WORK OF SOLICITORS AND BARRISTERS

Evaluate the different stages of the evolution of a local plan document for an area with which you are familiar. Identify and list the influences which may have affected its development from the draft stage. Consider what is:

- government policy or advice;
- local policy or political preference;
- influenced by local or national/international landowners and business;
- influenced by community, action groups, etc.

WHAT ARE THE OUTCOMES OF PUBLIC POLICY?

The outcomes of public policy may vary widely, both over time and over space. They may be broadly in accordance with the expectations of the policy or alternatively be the complete opposite. In addition there may be negligible impact of public policy, particularly if it is not backed up by statute or by resource commitments, such as funding or staffing.

Some policy may not work because of resistance to it in the marketplace. This may apply particularly to certain fiscal initiatives, such as parts of the Development Land Act of the 1970s, where certain sections of the community (landowners and developers) felt they would lose out substantially from the government's attempts to recoup a share of 'betterment value'.

Other policy can succeed, even with limited resources, because it captures the mood or opportunity of a particular period. Perhaps the greatest UK planning example was the pioneering Town and Country Planning Act 1947 when, as part of a new culture of post-war reconstruction, a British population accepted the nationalization of their development rights – something which proved harder to achieve in some other countries and would probably be difficult if tried again in the UK today.

If the population is behind a policy measure, it can make a difference to its longevity on the statute book. Duncan Sandys is widely acknowledged as the pioneer who introduced the first Conservation Area legislation in the UK. The fact that so many communities were in distress over the destruction created by post-war redevelopment, not only of landmark buildings like Euston Station but also of familiar, modest neighbourhood environments, helped to give this element of public policy greater public profile and staying power. It now has a similar weight and publicly perceived legitimacy as the older Green Belt concept.

The reality of public policy is that it exists within a complex interacting web of society's institutions and individuals. Policy analysts and civil servants, no matter how skilled, cannot always predict – and certainly cannot control – the behavioural outcomes of policy. Ultimately we must be thankful for this non-systematic behaviour of society, although it can make public policy formulation a somewhat precarious profession.

It is worth addressing why one should consult people in the development of public policy, particularly as it applies in the planning process [6]. Three reasons are suggested, although they are not definitive:

- because it is intrinsically fair and proper in a democratic system; part of the checks and balances which ensure there is no unchecked 'dictatorship' by those in power;
- to give greater legitimacy to the policies, plans and projects which form the basis of consultation, usually prepared 'in the public interest' using public money raised from taxes;
- to encourage greater efficacy of the policies, plans and projects through building up public awareness and support, and minimizing opposition to their implementation.

Different levels of consultation and involvement have been identified by writers and commentators in public participation. Arguably the best known is Sherry Arnstein who, writing in 1969 on the problems of immigrant and poor communities in American cities, identified a 'ladder of citizen participation' [7], shown in Figure 7.1.

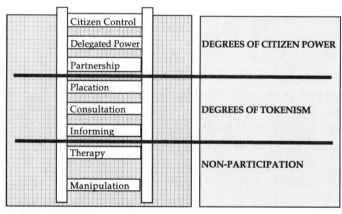

from Arnstein, 1969

Figure 7.1 Arnstein's Ladder of Participation. (Redrawn by Dick Pratt.)

The ladder is reasonably self-explanatory. Although others have evolved and adapted it, the main principle is that there is a conceptual graduation along a continuum from non-involvement, through information exchange, to fuller participation in the decision-making process.

This helps us to locate mentally different types of approach to the involvement of people in the formulation or implementation of public policy. The precise mechanism is discussed more fully in Chapter 8.

WHO ARE THE KEY PLAYERS?

It is worth exploring the role and motivations of some of the key players in the process of public policy, focusing on the example of operating the planning system locally.

THE PLANNERS

In the UK planners are trained professionals who operate a statutory planning system through local planning authorities. They may be broadly divided into two categories:

- the policy-makers, who formulate plans, proposals, policies and strategies;
- the plan implementers, who administer planning and development policy, most notably through determining applications for planning permissions or licences.

Whilst planners can be both policy-makers and implementers at the same time, and junior planners can experience both in their training, the two roles are often divided by aptitude and interest for much of the professional life of most planners.

The main aim of the policy-makers is to devise plans and policies and thereby influence development and even possibly wider quality of life issues of the citizens of their area. They are frequently concerned with matters of research, data, national policy and future trends. Their 'currency' can become policies, plans and strategic visions and they can – deliberately or otherwise – become divorced from the implementation aspects of their own output, particularly if the policy becomes perceived as an end in itself rather than a means.

Professional policy-makers frequently have – or need to have – strong relationships with politicians who may be, through their political manifestos, the originators of the broad policy directions. Where this is not the case, it may well be that professional policy-makers generate policy ideas and initiatives and seek to persuade politicians of the relevance and importance of their output.

In some instances policy-makers may draw upon public opinion to help devise their policy, although it is more likely that they seek opinion once they have devised policy ideas themselves, or in consultation with politicians. The strength of policy-makers is that, through their concern with data on longer-term trends and strategic issues, and a culture which builds up around this, they can take a broad long-term view of a particular matter of concern. Unfortunately, the weakness of such a position can be an occasional lack of empathy with the community upon whom any policies and proposals will impact on a day-to-day basis; that is, a professional sense of detachment (or even alienation) from the effects on everyday life and everyday people.

It is important to identify and understand the role of policy implementers in this respect, because they may have to 'sell' the public policy to other members of the public in their dealings with them. The role of policy implementer can therefore breed a slightly different attitude, particularly in the function of development control, where there is less of a long-term strategic vision and more of a short-term pragmatism born of dealing with a more intense pressure from a range of individuals and organizations. These might include pressures from politicians or developers (for instance, to ignore or override plan policies) or from members of the public (say, to prevent a development). The day-to-day currency by which performance is gauged can become reduced to rapid decision making on recommendations, or deal making in negotiations. There is probably greater potential for face-to-face conflict in this role, more contact with the 'outside' public, and the need for slightly different social and professional skills. A different psyche is necessary to perform this interfacing, mediating role.

WORKPIECE 7.3

THE RESPONSIBILITIES OF POLICY-MAKING

You are a planner in a local authority. You have previous training in both policy formulation and implementation aspects. List the reasons you might be motivated to become a planner and your main areas of interest. Identify what you would believe to be the enjoyable and problem areas with each of the policy formulation and implementation roles. What difficult situations might you encounter in the workplace? Which would you prefer to deal with? Why? How can you make the difficult role more enjoyable and more effective?

THE POLITICIANS

Planners will frequently say it is the politicians who make policy and decide. This is true where elected politicians have a discretionary role. However, in many countries a plan – once it is adopted – may be simply administered by officials.

Politicians can be important at two levels: national and local – at the level of a planning authority. Generally national politicians stay out of planning matters, unless:

- they are particularly interested themselves, or have a key responsibility in the field;
- intensity of public interest forces them to become involved because there may ultimately be votes at stake.

They tend only to become involved in large or controversial projects. As national politicians, they may well see professional planners as merely local bureaucrats, and their prime concern is their constituents. They may well seek to override planning policy in some of these controversial areas, particularly if it is electorally advantageous.

Local politicians frequently have a stronger interest in the planning system because:

- it has a close relationship with and impact upon the local community who elect these politicians;
- it can assist in the generation of development, jobs, civic pride and prestige, all of which may be motivating factors for local politicians;
- it can lead to the downfall of politicians or parties if badly handled, in a way that is often less likely for national politicians.

Therefore, local politicians frequently have a strong interest in the planning system. It gives them a genuine role in shaping the future of their community. Ironically in Britain, because the nature of the system is discretionary and the motivation of politicians is usually to exercise power in decision making on behalf of the community, they may tend towards a preference for decisions to be made on an *ad hoc* basis rather than on the policies and provisions of a statutory development plan. Whilst it is true that the relative significance of adopted plans and policies may vary, it is frequently the case that politicians do not like to see their discretionary powers converted into 'rubber stamping' an approved plan. What would be the fun in that?

The strength of politicians, particularly good ones, is that they can have a wider social and economic perspective than professional planners and can (sometimes even have to) adapt to changing moods and priorities in the wider population. Planning is not their sole concern, as it can be with the professionals, and therefore planning is only one element in a range of public policy areas.

The weaknesses of politicians can range from a simplistic over-adherence to policy documents and a lack of breadth and lateral thought, to seeing planning purely as a mechanism to support initiatives in their own constituency (e.g. major investment in a leisure centre), when a broader perspective is needed. Perhaps the greatest danger can occur when politicians ignore their own plans and policies and guide the development of their area through a series of *ad hoc* decisions. This does not give the local community, traders, business or developers any clear views as to their intentions, nor does it ensure consistency and fairness in the implementation of public policy.

The public who interface in some way with the planning system are its general consumers. Two broad categories may be identified:

THE PUBLIC

- those with an active operational or financial interest in the plan or policy (**vested interest groups**);
- those who only have a broad interest as 'consumers' of the plan and local environment (**passive consumers**).

These are somewhat simplistic distinctions, not least because individuals may fall into both categories. The vested interest groups might include local traders and business people, or landowners and developers. Passive consumers might include local residents, schoolchildren, road users, shoppers, leisure activists and employees.

The motivation of the vested interest groups will probably be to secure the value of their investment and/or the smooth running of their business into the future. For landowners, for instance, this may entail making strong representation to the planning authority to prevent over-restrictive land use policies or notations applying to their land. For a business it might involve lobbying to safeguard additional expansion space for their operation or even to object to a competing business operation (e.g. retail foodstore).

The motivation of passive consumers is less likely to include seeking personal gain, although this can apply through, say, seeking special protective designations – for example, conservation areas. It is more likely to form an avoidance of loss, perhaps of an open space, a cherished building or a community facility. In addition, concerns may arise about long-term degradation of the environment (air pollution, land or river contamination) and the effects on community health.

In the UK people have generally been able to get involved at three basic stages of public policy making in the planning process: in development plan making; in development control; and in special projects and proposals. These stages can be found in many systems. How do they compare with your own system? The issues which may influence the appraisal of site and area potential are illustrated in Figure 7.2.

**WHERE CAN PEOPLE
GET INVOLVED?**

165

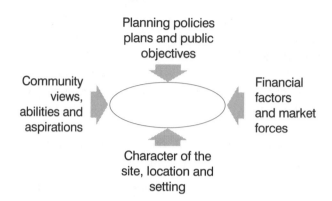

Figure 7.2 Issues for site and area appraisal. (Developed from a diagram by R. Silson.)

DEVELOPMENT PLAN MAKING

Individuals, organizations and community groups can engage in the policy-making process through the opportunity to comment on a draft plan and its revised versions. In some instances, communities or representative groups have played a part in feeding ideas and proposals into the process before the plan is produced, although this normally depends on the openness of the authority. If bodies remain opposed to aspects of a plan even after it has been modified to take account of observations, it may be possible for them to be heard at a public inquiry into the objections.

In the UK this is a significant improvement on the minimal involvement previously available but it has a number of weaknesses. The process tends to be directed at specific groups, organizations and landowners, and is less accessible and comprehensible to ordinary members of the public. Much of the effort and scrutiny is devoted to the objections (often from vested interests such as landowners) and less into the overall quality, depth and community support for the plan. The plan is therefore very much a public policy document of the planning authority and not of the public.

WORKPIECE 7.4

DEVISING A CONSULTATION PROGRAMME

You are a planning professional in charge of developing public policy and an action plan for the revitalization of a medium size town centre. You need to consult local people. Devise a consultation programme, including advanced publicity to maximize attendance and the identification of a suitable venue. Decide:

- what the purpose of the meeting is;

- who should be contacted or specifically invited;
- how to publicize the event;
- what will be said at the meeting;
- how the comments will be fed into the policy generation stage;
- what the follow-up steps are to keep the public informed.

The development control process is the mechanism whereby applications made for development are assessed against the provisions of the development plan and other policy considerations, and a determination is reached by a planning authority. This is usually either refusal or permission (the latter normally with conditions). The terminology can vary, depending on whether an application is for a building, for signs or for works to a listed building, for instance, but the basic types of determination are similar in many countries. If an application for permission is refused, the applicant usually has a right to appeal.

In general, the public can be made aware of applications through a variety of mechanisms such as newspaper notices, published application lists and letters to adjoining owners. The precise method varies according to the specific legislation and the approach of different places. This participation by the public in the process of implementing policy is important. It is in effect a feedback loop which can influence the formulation of future policy, particularly if public opinion is strongly expressed against the existing policy. Is there effective participation where you live?

Relatively few individual members of the public make representations about individual applications. Far more public consultation occurs via established groups, such as civic societies or neighbourhood action groups. These are often held to be representative of public views even though their memberships are often not statistically representative [8]. But they do form a useful extra channel of communication.

One of the difficulties of development plan and development control is translating broad policies with physical form. Urban design strategies may offer the means of bridging that gap and Figure 7.3 illustrates some of the potential.

DEVELOPMENT CONTROL

Increasingly new ways have been devised to engage and involve the community not only in the process of evolving public policy but also of implementing it, or at least aspects of it. Some of the most innovative methods have been developed, perhaps necessarily so, in the most problematic, peripheral communities. In all probability more middle class, articulate communities have various other routes to bring influence to bear on planning authorities, for instance by opposing road schemes.

These community mechanisms have included forming specialist action groups, neighbourhood alliances to have greater impact on decision makers and even large-scale mass action such as the opposition to the UK Poll Tax and Criminal Justice Act 1995, and the student unrest in North America and Europe during the 1960s and 1970s.

SPECIAL PROJECTS AND POLICY INITIATIVES

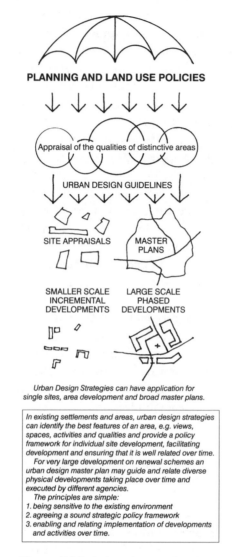

PLANNING AND LAND USE POLICIES

Appraisal of the qualities of distinctive areas

URBAN DESIGN GUIDELINES

SITE APPRAISALS   MASTER PLANS

SMALLER SCALE INCREMENTAL DEVELOPMENTS   LARGE SCALE PHASED DEVELOPMENTS

Urban Design Strategies can have application for single sites, area development and broad master plans.

In existing settlements and areas, urban design strategies can identify the best features of an area, e.g. views, spaces, activities and qualities and provide a policy framework for individual site development, facilitating development and ensuring that it is well related over time.
  For very large development on renewal schemes an urban design master plan may guide and relate diverse physical developments taking place over time and executed by different agencies.
  The principles are simple:
1. being sensitive to the existing environment
2. agreeing a sound strategic policy framework
3. enabling and relating implementation of developments and activities over time.

Figure 7.3 Relating plans to site development (David Chapman).

Many innovations have occurred in peripheral social housing areas such as Easterhouse in Glasgow or Hulme in Manchester, or in mixed communities in Northern Ireland. Initiatives have included community trusts and companies to stimulate social and economic as well as physical renewal. Planning-for-real and regeneration workshops have been used to foster a sense of community ownership and commitment to an action programme for the improvement of an area.

Increasingly these community development techniques have been more widely used by local authorities, development corporations and organizations such as the Civic Trust as tools for all kinds of public involvement in the process of policy formation and renewal. The public-transport biased 'All Change' transport policy for Scotland's Central Region was developed using community and business workshops, as was the strategy for upgrading Edinburgh's Royal Mile.

These kinds of innovative 'bottom up' techniques can only work where the community is given a genuine participatory role and have tended to fall apart if either politicians or local government officers have too much of a 'top down' control-driven culture, where consultation is merely information exchange, or where the community is irrevocably divided.

A collaborative approach
• **V** ision
• **O** wnership
• **C** ommitment
• **A** ction
• **L** ooking after it

Figure 7.4 A checklist and agenda for collaboration in urban design. (Developed by Drew Mackie, Tibbalds Munro.)

WORKPIECE 7.5

ASSESSING PUBLIC OPINION

An application has been received by your planning authority for permission to construct a private heliport to support the likely future growth of a nearby offshore oil industry. You have no specific policy on heliports, although you have policies to support local economic development and industry, and to protect the environment and residential amenity.

The heliport would generate frequent loud noise along a corridor. Residents are angry and apprehensive about its likely impact. Arrange a suitably scientific assessment of public opinion (e.g. a questionnaire) and 'role play' a public meeting by acting out the roles of the interested parties, including:

• helicopter company (applicant);

• oil companies (users);
• local residents;
• local Chamber of Trade and business;
• local environment group;
• local planners and environmental health officer.

Balance the different views on the issue and recommend whether the proposal should proceed or not.

This new collaborative approach may be symbolized by the acronym VOCAL, which was developed by one of the community consultancy pioneers, Drew Mackie, when working for Tibbalds Monro. The checklist (Figure 7.4) runs through from a shared vision, arrived at consensually, to commitment, action and aftercare, which can only occur if individuals feel some sense of ownership.

SUMMARY

Public policies set the economic and legal framework for the activities of institutions and individuals. Importantly these policies guide priorities and action. Strategic and local plans, combined with service delivery and incentive schemes, both guide and encourage improvement in the quality of the built environment and the way it meets people's needs.

In many areas of modern political and social life, individuals and communities expect to have a say. Indeed, in many societies they have specific rights, enshrined for instance in a Bill of Rights. This was not always the case, nor was the consultation of the public necessarily an integral part of the land use planning systems which evolved in this century.

In those western states where planning now embraces some form of public involvement, it was largely won against a background of resistance or benign neglect. For instance, in Great Britain there was no real role for the public in the planning system until the Town and Country Planning Act 1947. After that momentous Act individuals gained the right to object to development plans and the proposals they contained, although this fell short of full participation.

As comprehensive development and renewal proceeded apace during the 1950s and 1960s there was a groundswell of public opinion against such things as motorways, new offices and shopping centres and loss of well loved historic areas. In Britain this was often an expression of the growing conservation lobby against the loss of familiar public buildings such as the nineteenth century Euston Arch, but it was also a reaction against the break-up of traditional communities. People felt excluded by the policy and decision-making processes and ignored by those in power who pursued the path of 'progress and technology'.

Against a background of growing civil unrest in many parts of the world, and much community objection to development in the UK specifically, a report concerning public participation in planning was commissioned in the late 1960s. The 1969 Skeffington Report, as it is widely called, sought to introduce new mechanisms whereby the public could become engaged more constructively in the planning process. It was influential in as much as it set the tone for the level of public participation in the statutory planning system in the UK over 20 years.

CHECKLIST

This chapter has considered:

- the purposes and forms of public policies, planning laws and practice;
- why people should be consulted about both policy and implementation;
- the roles and relationships between some of the key players in the process of public administration ;
- how the legal system and frameworks of plans relate to implementation, guidance and control;
- some of the obstacles to the achievement of public policies and full participation in decision-making processes.

REFERENCES

1. Bruton, M.J. (ed.) (1984) *The Spirit and Purpose of Planning*, 2nd edn, Hutchinson, London.
2. Cullingworth, J.B. and Nadin, V. (1994) *Town and Country Planning in Britain*, 11th edn, Routledge, London.
3. Devas, N. and Rakodi, C. (eds) (1993) *Managing Fast Growing Cities*, Longmans, Harlow.
4. Greed, C. (1993) *Introducing Town Planning*, Longman, Harlow.
5. Rydin, Y. (1993) *The British Planning System: An Introduction*, Macmillan, London.
6. Department of the Environment (1969) *People and Planning: The Skeffington Report*, HMSO, London.
7. Arnstein, S. (1969) A Ladder of Citizen Participation. *Journal of the American Institute of Planners*, **35**(4).
8. Lowe, P. and Goyder, J. (1983) *Environmental Groups in Politics*, Allen and Unwin, London.

FURTHER READING

Ambrose, P. and Colenutt, R. (1975) *The Property Machine*, Penguin, Harmondsworth.

Cherry, G. (1982) *The Politics of Town Planning*, Longman, New York.

MacArthur, A. (1993) Community partnership: a formula for neighbourhood regeneration in the 1990s. *Community Development Journal*, **28**(4).

Mackie, A. (1992) *Good Practice Advice Note: Community Involvement in Regeneration*, Civic Trust Regeneration Unit, London.

Ravetz, A. (1960) *The Government of Space – Town Planning in Modern Society*, Faber and Faber, London.

Stoker, G. (1988) *The Politics of Local Government*, Macmillan, Hampshire.

Thornley, A. (1991) *Urban Planning Under Thatcherism – The Challenge of the Market*, Routledge, London.

Wates, N. (1976) *The Battle for Tolmers Square*, Routledge and Kegan Paul, London.

COMMUNITY ACTION AND INVOLVEMENT

DICK PRATT

THEME

In some localities, there is a higher level of community involvement in the use and development of the area. In others, the level of involvement and participation is low and contributes to feelings of alienation and hostility. This chapter explores the opportunities and need for positive community and individual involvement and action. It identifies some of the obstacles to participation, steps which can be promoted by public policy and ways in which communities can exert influence from the 'grass roots'.

OBJECTIVES

After reading this chapter you should be able to:

● understand better the factors which lead to greater or less public involvement in the development or redevelopment of their localities;

● recognize different types and levels of public participation through consideration of some case studies;

● examine cases of public participation and suggest reasons for the differences encountered;

● understand the complexity of factors determining whether a chosen method of participation is deemed satisfactory.

INTRODUCTION

Social historians are very keen to explain why the issue of public participation and involvement waxes and wanes according to long-term, large-scale social processes. Sociologists too are interested in the ability of social systems to deal with varying levels of demand for public consultation. They point to the importance of involvement of people in establishing a stake and an interest in the outcome of changes to the built environment. In this their focus of attention is shared by political scientists. Why should the public be involved in planning and development processes? Opinions will vary as to the overall significance of involving the community. For example, Crow and Allan [1] have commented that:

> the attraction of community involvement within contemporary policy making is less about democratic self determination and more about managing social tensions and assisting state bureaucracies to accomplish their objectives.

This idea can be summarized as expressing the tension between incorporation and self-determination and accompanies most political processes. Arnstein [2] set up a continuum of citizen participation that stretched from token involvement of the public through to full blown citizen power (Figure 7.1). How else can the issue of citizen participation be viewed? In particular what policy perspectives can inform the fullest type of participation exercise? A recent UK report *Community Involvement in Planning and the Development Processes* [3] provides cogent guidance. Here the authors describe an approach where 'involvement is conceived broadly and considered integral to planning and development; demanding open ended approaches and generating added value'.

An important distinction needs to be made between the involvement of the public in the plan-making process and public involvement in a specific development proposal. In the UK the Skeffington Report [4], published in 1969, covered both aspects. In practice the former has remained relatively undeveloped in the period since then, though recent legislation has laid down minimum periods in which the public can express opinions at draft and final stages [5].

It is important to reflect upon the general issue of democracy and the built environment. More than any other, the twentieth century was the battleground of ideas about the forms of representation that should be adopted. Industrializing countries emerged from the technological revolutions that provided the material basis for wealth and prosperity and brought with them a variety of 'modernizing' ideologies. In these countries political orders based upon theology largely gave way to liberal democracy, communism or fascism. The power of authoritarian state socialism from the left and totalitarian nationalism from the right has for the moment been exhausted. It is only just possible to imagine a resurgence of such movements from time to time under extreme social and economic crises.

NEW POLITICS – NEW AGE?

Religious movements of both the developing and the developed countries have increasing influence in many places. At the same time 'rightist' and 'leftist' versions of liberal democracy continue to have important effects upon the interpretation and implementation of state policy, including how the question of public participation in decisions about the shaping and management of the built environment should be handled. An added political dimension is that which has been established by the 'green' movement. It has had a very significant influence on the way in which governments, opposition parties, administrative bureaucracies and business think and act.

All these ideological trends can be represented very simply in terms of attitudes towards ourselves (the insiders) and themselves (the outsiders), towards 'them' and 'us' (Figure 8.1). This needs to be analysed at two levels: the macro and the micro. We all live in a world where the nature of the built environment that we have locally will have impacts upon the world in general. We clearly need to be able to understand as well as possible how the micro-culture of the locality will articulate with the macro- culture regionally, nationally and internationally.

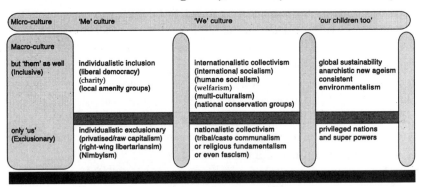

Micro-culture	'Me' culture	'We' culture	'our children too'
Macro-culture			
but 'them' as well (Inclusive)	individualistic inclusion (liberal democracy) (charity) (local amenity groups)	internationalistic collectivism (international socialism) (humane socialism) (welfarism) (multi-culturalism) (national conservation groups)	global sustainability anarchistic new ageism consistent environmentalism
only 'us' (Exclusionary)	individualistic exclusionary (privatised/raw capitalism) (right-wing libertariansim) (Nimbyism)	nationalistic collectivism (tribal/caste communalism or religious fundamentalism or even fascism)	privileged nations and super powers

Figure 8.1 Collectivist/individualist grids and the basis for local public involvement. (Dick Pratt.)

In very general terms it is possible to divide the conditions under which public involvement in property development is important into two broad categories. Firstly there have been an increasing number of protests in recent years as a consequence of proposed land takes for new developments. These can include proposals for new roads, housing and retail developments. In the case of retail developments or airports, there may be very significant impact upon existing settlements. The second general category is that of the redevelopment of the existing built-up areas. Such redevelopments have often occurred as the result of the obsolescence of existing land uses – the closure of a steel mill, the abandonment of factory buildings or hospital lands, the dereliction of docks, railway lands or gas works. The most intense forms of public consultation have usually accompanied proposals to redevelop obsolete forms of residential land.

The UK's comprehensive redevelopment programme of the 1950s and 1960s was most often criticized for the absence of adequate public consultation [6]. In contrast, the remodelling of the modernist UK housing estates of those two decades has been subjected to the most intense forms of community consultation and participation in the 1980s and early 1990s. High-rise flats are the most visible but far from the most numerous housing form of the 1960s. Together with their accompanying large, featureless open spaces between the blocks, they have become very unpopular. In reshaping these spaces important lessons have been learnt in the UK about the need to involve local people through initiatives such as Estates Action, Priority Estate projects, Safer Neighbourhood projects and other neighbourhood initiatives. Under conditions of liberal democracy, the ability of ordinary people to make their views felt varies enormously. The views of local people may be informed or mediated by a variety of different political ideologies – conservationist, environmentalist, liberal, socialist, anarchist, religious, cultural or feminist.

The changes which have occurred to the economies of the industrialized world have been so great that some commentators have argued that we are living through an era of post-industrialism. One characteristic of this epoch is said to be the uncertainty and fragmentation of the political process and the way this affects how power is distributed and struggled for.

POWER

Liberal democracy is a system where people have an inalienable right to be consulted over the type of government on a periodic basis. Many aspects of the management of the built environment have become subject to these democratic principles. This is partly as a result of the need to justify or legitimate the activity of the state in taxing people and spending the proceeds. Historians of the nineteenth and early twentieth centuries portray a situation of gradually improving urban conditions. As a result, the political struggle of the urban masses to exercise more influence over the management of their environment became part of a larger picture to disperse power more widely in society.

Despite formal equality of access to decisions in law about the built environment for all citizens, it is clear that some groups can exercise more powerful influence on the outcome of decisions than others. This discrepancy arises from two sources: firstly, from ownerships of land assets; secondly, from possession of knowledge and skills.

Ownership of land assets bestows an immediate advantage because international law generally favours those who own land and want to do something with it. Of course there are notable exceptions. The UK land use control system tends to 'freeze' existing uses and intensities of use. So, for example, a farmer who becomes tired of farming and seeks to retire to a life of luxury would not be able to sell his fields for luxury homes, supermarkets or a business park without a serious change to the relevant local or structure plan.

SKILLS AND KNOWLEDGE

The possession of knowledge and skills gives some people a particular advantage. If a local authority planner concerned with handling objections to development proposals is asked which areas excite most interest, the answer will generally be those areas in which reside people who have had the opportunity of post-school education. They are more confident and eloquent when it comes to the processes which have been installed to handle objections. This involves telephoning planning officers, writing letters, maybe organizing petitions or even protest meetings. At the beginning of the 1970s the recommendations of the Skeffington Committee on public participation in UK planning began to be implemented and the framework has empowered these relatively better educated people to voice their opinions. However, the majority of the population who are affected by development proposals do not even know where to begin when it comes to understanding the impact of a development, let alone effectively registering an objection to it. Similar conditions can be found world-wide. How does your area or place of study compare?

Hence the emphasis of commentators writing about public involvement today is one still based upon social class. Generally middle classes and the professional groups appear to have excellent access to local political élite who make important decisions, while the working class are involved only episodically with the management of their environment. Other factors are also important – for example, the extent to which a community is settled or the extent to which it is subject to a high turnover of residents. This applies particularly to developing countries where population growth and migration create dynamic settlement patterns, but the effects can equally be found in inner-city areas world-wide.

IDENTIFYING LOCAL ISSUES

Ask people in your own locality what planning and development issues have most affected them in recent years. Ask about roads, traffic calming, parking, buildings, parks, open spaces, footpaths, safety at night, etc., using language that they will understand. Produce a summary report on the issues you have discovered.

HOMES AND JOBS

Fundamental changes have occurred in many labour and housing markets in the last 20 years. Large-scale, stable employment for most men has shrunk drastically; female employment has increased but on the basis of precarious part-time, short-term patterns. Adolescents contemplate the prospect of spending at least some years unemployed or in the various job experience schemes. Those in employment feel obliged to work for longer hours to safeguard their labour market position. They are working longer hours not primarily to boost their disposable income but to hang on to employment itself.

The homes that people live in came to be seen in a very different light in the UK during the 1980s. For groups who had the prospect of improving their standard of living, their residences were perceived as assets rather than merely as homes. This increased the amount of local mobility and people jostled in the housing market for a superior asset. These factors loosened the ties of community and locality in many different areas of the UK. The house price slump of the late 1980s and the negative equity of some homes compounds the problems for many people and communities.

COMMUNITY OR NETWORK?

The mass organization of communities described so eloquently by the community studies of the 1950s and 1960s [7] has all but disappeared. The gradual erosion of industrial communities based upon a similar set of relationships to the means of economic production has left many working-class localities bereft of a single employment basis and during the last 15–20 years all types of collective spirit have been severely strained. Increased personal mobility through wider access to private car use and through educational qualifications has led to the stretching of the spatial ties of family and kinship.

The result of these trends is that fewer people have a stake in any one single location. Fewer assertive, mature and responsible adults are available during daylight hours to keep an eye on the neighbourhood. Those in work have the resources but not the time for community involvement. Those out of work have the time but not the resources to be engaged in community tasks. It has become fashionable to use the term 'network' to describe the interconnectedness that people maintain through work, mutual interests or historical connections. Whereas networks will fulfil many human needs for sociability and mutuality, they singularly fail to 'fix' human relations around particular places in the way that 'community' was claimed to have done.

COMMUNITY AND LOCALITY

Planning policies have sometimes exacerbated the spatial divisions between employment and residency. The baseline of community spirit of middle-class districts has been reduced to the most elemental NIMBYism (Figure 8.1). These 'not-in-my-backyard' responses may defend individuals' environmental assets, but they are incapable of advancing the environmental common good. Nevertheless, new forms of collective and community organization have developed and evolved and new definitions of community have come into existence. At the same time politicians appeal to a 'sense of community' to justify in rhetorical fashion traditional values which can win popular support. This can be more easily produced than the practical answers to the difficult problems that are faced.

New forms of community organizations have appeared during these last 20 years. The surge in interest rates on borrowed money and increasing community indebtedness produced credit unions during the late 1980s. These have provided important 'glue' helping to hold working-class communities together, a function formerly performed by trade unions and tenants' organizations. These associations have fragmented in the face of large-scale changes to the structure of labour and housing markets, but others have emerged.

Local exchange trading schemes (LETS) encourage mutually beneficial economic exchanges, often using informal currencies. So far about 200 are registered in the UK with about 10 000 people using the alternative currency cheque books [8]. Urban wildlife groups captured the imagination of community members who would have been otherwise forced into idleness as a result of employment redundancy and premature retirement.

In the UK, middle-class communities faced with the threat of road building or road widening pitched into battle, developing a range of new methods of struggle, utilizing the public consultation methods introduced under the Skeffington reforms. Working-class communities appear not to have availed themselves of the mechanics of public participation in the planning system. The round of public consultation involved in plan making at the end of the 1980s and early 1990s largely passed the working-class communities by.

In the following sections you will encounter some case studies that illustrate a variety of elements of public participation and involvement (Figure 8.2). These are set out in the framework in Figure 8.3.

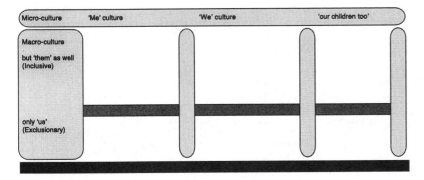

Figure 8.2 Example of collectivist/individualist organizations acting with local public involvement. (Dick Pratt.)

ASSESSING ORGANIZATIONAL POSITIONS

Having read the introductory argument and considered Figure 8.1, identify some examples of the various kinds of organizations which are reported upon in your national or local newspaper. Locate each of these examples in a corresponding part of Figure 8.3.

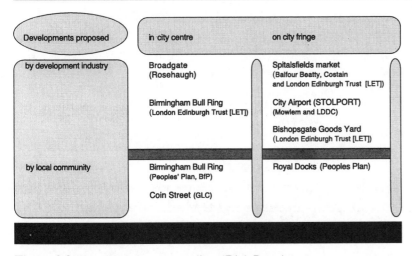

Figure 8.3 Framework for case studies. (Dick Pratt.)

179

EXAMPLE 8.1

LONDON DOCKLANDS

Docklands does not really exist as a single local community – it is made up of sections of several London boroughs and takes in parts of diverse traditional community areas such as Limehouse, Wapping, the Isle of Dogs and North Woolwich (Figure 8.4).

In Docklands during the early 1980s, local government discovered forms of community consultation that empowered rather than disenfranchised working-class communities. However the Greater London Council's efforts at popular planning in Docklands came too late to be realized. The Conservative Government had lost patience with the apparently slow rate of redevelopment achieved by Labour-controlled Docklands boroughs and had imposed its own development quango, the London Docklands Development Corporation (LDDC) [9–12]. Although the views of local people had been formulated and documented, these were largely ignored by the new quango, as Brownill [9] points out:

Strategies designed to meet local needs were seen as old-fashioned and constraining on the activities of capital.

For example, in response to calls to match new jobs to the skill of local people, LDDC answered in 1992:

This represents a very limited horizon which would constrain the marketing strategy and the new Docklands economy that is required to provide a secure future for the area.[13]

Alternatives to the market-preferred options had been elaborated at an earlier stage. Brownill cites the People's Plan for the Royal Docks, the Limehouse Petition and the Docklands Child Care Project [9]. These became the rallying points for local people and ultimately forced from the Development Corporation a more community-oriented style, not only for the LDDC but for the later generations of urban development corporations.

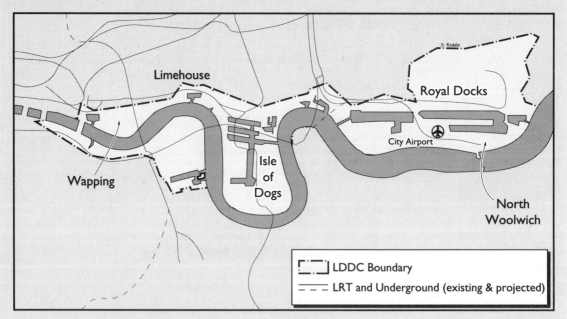

Figure 8.4 Map of London Docklands.

The People's Plan for the Royal Docks envisaged the return of ship repair, but such a goal would remain unrealizable without the raft of central government industry and trade policies which could make it possible. The People's Plan for the Royal Docks was formulated as an alternative to the City (Short Takeoff and Landing) Airport. This had been proposed by a development company (Mowlem) and the LDDC. As such it was connecting the area to business in Europe and the City of London, but doing so in a way which would barely touch on giving positive benefit to local people, who would instead have to put up with a worsening of the environment. The next example also takes up the theme of connection to the global economy.

EXAMPLE 8.2

SPITALFIELDS AND THE CITY OF LONDON FRINGE

Office space in the north-east corner of the City of London attracted a great deal of interest from corporate investors during the 1980s (Figure 8.5). The area around Broadgate saw several significant developments in this period – London Wall (Alban Gate), the European Bank in Bishopsgate, Exchange House over the railway at Liverpool Street station, and Broadgate itself. The Broadgate development provided the most sound investment in what proved to be an extremely risky form of speculative office-based development boom [14]. The relatively lower land costs, the possibilities of creating air-rights buildings, which use the space over pre-existing land uses such as roads (e.g. Alban Gate) or railways (e.g. Exchange House), lower cost car parking, proximity of train stations, lower cost food and other services plus the allure of real street life gave the location the edge for firms looking for floor space.

Spitalfields was a similarly promising location. Land values were low, the people were poor and without power. Yet it was near enough to the heart of the finance business – the 'golden' square mile of the City of London.

In response to this challenge to make money on the city fringe, the East London Partnership was formed representing 45 large and medium sized firms across East London. A consortium consisting of elements of Balfour Beatty, Costain and London Edinburgh Trust put forward a scheme for offices and spent £40 million moving the wholesale market to Hackney [15]. The London Borough of Tower Hamlets held the power to grant planning consent. However, in 1990 the Secretary of State for the Environment, responding to the conservationist

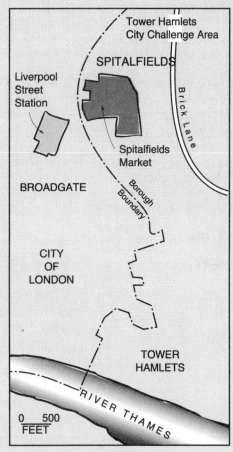

Figure 8.5 Spitalfields and the City of London Fringe.

lobby of English Heritage and the Royal Fine Arts Commission, called in the application and a decision was delayed. The over-supply of office space in the meantime meant that the development would no longer give a worthwhile rate of return for investors. Community opposition to the redevelopment continued throughout this period [15].

Residential gentrification of urban streets has been a comparatively rare occurrence in Britain in the recent past but it became a real issue for the area between Brick Lane and Bishopsgate [16]. The area's proximity to such well-paid jobs in the financial services sector initiated changes to the social mix in the Brick Lane area. Community activism had arisen in the area from the need to relieve housing overcrowding and to resist racist attacks [17]. This does not mean that the community was or is united [16] and indeed the divisions have led to disagreements in the face of property proposals for the area [15].

In addition to Spitalfields Market, two other notable schemes were proposed for sites in the London Borough of Tower Hamlets – sites which would have derived form, function and value from proximity to the City Fringe. These were the Truman's Brewery in Brick Lane and the Bishopsgate Goods Yard. In the case of the latter the London and Edinburgh Trust (LET) felt obliged to engage in a lengthy public participation exercise carried out by community architect firm, Hunt Thompson, to secure a modicum of support for a scheme of mixed uses including some offices [18].

The Bishopsgate Goods Yard consultation was done by a series of 'planning weekends' and 'planning for real' exercises. This resulted in a scheme in which progression through the elongated site from west to east shifted the emphasis from offices to social housing and community workspaces. Promise of this kind remains to be realized. The property tide ebbed and Tower Hamlets has had to look to the competitive bidding system for urban aid to begin to realize comprehensive area-based urban regeneration. The Spitalfields Community Development Group argued that the two sites should be included in a plan for the whole area, which would include a 'Banglatown' retail centre and demanded housing and training before offices [15]. Tower Hamlets made a bid for central government funded grant aid, under the title of 'City Challenge', for Bethnal Green on the edge of Broadgate and Spitalfields. The grant was awarded in 1992. This urban regeneration initiative is pioneering a variety of types of community involvement in terms of developing the capacities of local people. Managing this process so that people with less formal status in the community are not 'crowded out' is one of its important challenges [19].

EBB AND FLOW OF INVOLVEMENT

In modern societies, effective and stable long-term involvement by local people requires a firm commitment to that goal by the central state. Where the central state is in the least part indifferent, local involvement finds itself expressed as opposition. But where the central state has a more positive attitude, this quality of public involvement could be one of several kinds.

Community participation in projects in the UK has received an ambivalent response from central government during the last 15–20 years. By contrast the land use plan-making and development control system has been more or less continuously subject to the requirements of the Skeffington principles established over 20 years ago. As noted earlier, the middle classes and professional groups make considerable use of rights afforded under these principles. By contrast, community participation in working-class communities may depend on the discretion of the legally constituted development agency. However, there are opportunities for this to change with a wider recognition of the way that community involvement can add value to a scheme[3].

For a brief period in the UK during the 1980s there was a 'presumption in favour of development'. This meant that although land use plans had gone through a public consultation process they could still be overturned if a developer managed to convince the authorities that a development was worthwhile. However, this lead to an over-supply of speculative development in many places, and since the end of the 1980s the 'presumption' has been in favour of the 'plan'.

The skill of objectors, particularly where the knowledge of experts is wedded to the articulateness of the educated and the force of whole communities, has meant that major development schemes have almost ground to a halt. Recent well-known examples have been the cases of motorway and trunk road protests – for example, the battle for Twyford Down and the East London river crossing. Other popular campaigns have been concerned with urban redevelopments, such as the King's Cross railways lands and the battle of the Birmingham Bull Ring. A model for local participation for redevelopment was Coin Street, a project for homes and work places conceived in the early 1980s and eventually realized when the Greater London Council gave it support in the mid 1980s [20][21]. Coin Street had a dramatic impact upon the scale of the urban form in a part of London dominated by the headquarters of multinational companies, but it remains an isolated example.

For a long while,community participation in projects was out of favour with central government in the UK but in the mid 1990s this is once more changing. Central government has brought out new advice on community consultation. The guidance [3] assumes a wide diversity of situations and people to which participative approaches now apply and it assumes that community participation and involvement:

- improves the quality of developments (the 'added value' argument);
- helps to reduce potential conflict;
- helps to avoid adversarial conflicts and hence will speed up the pace of development and contribute to reducing costs.

It recognizes that community involvement will not always guarantee success, which itself depends upon good management of the process, realistic expectations and real consultation taking place – not just information being given.

OBJECTIONS AND OBJECTORS

A variety of 'actors' are identified in this report: developers, local authority planning officers, other government officers, elected members, consultants, community and voluntary groups, non-government organizations and community enablers. The report aims at achieving better practice and expresses the hope that this will be achieved through a better management of the whole process. It recognizes that not every public involvement exercise would even aim to scale the higher rungs of Arnstein's ladder [2]. It considers that such a process would need to be guided by a checklist composed of several stages (Example 8.3).

EXAMPLE 8.3

INVOLVING PEOPLE: A CHECKLIST (ABRIDGED FROM *COMMUNITY INVOLVEMENT IN PLANNING AND DEVELOPMENT PROCESSES*, DOE, 1994 [3])

- Initiating involvement: what are the factors, positive or negative, which provide the context for involvement?
- Communities of interest: can they be identified?
- Commitment: what commitment to collaboration do the various interests have?
- What methods can be employed to enable participation?
- Will the approach be appropriate and adaptable to changing circumstances?
- Can participants gain ownership of the process without dominating it?
- Can they share experience and learn to collaborate?
- Do they agree the scope of participation, or do some oppose or wish to limit it?
- Are any interests outside the process and could this constrain outcomes or involvement?
- How can the process be programmed from initiation to long-term monitoring?
- Can conflicts between interests or objectives be resolved or accommodated?
- How can we evaluate the experience and develop better practice for the future?

REGENERATION AND BUSINESS

In the UK the strategy for urban regeneration has been built around certain expectations:

- that larger businesses have some capacity for local urban regeneration activities;
- that smaller/medium sized firms have affinity but less capacity for urban regeneration;
- that all businesses need more support than merely improved access to land and buildings.

Despite having cut the programmes which benefited voluntary organizations, there is an awareness that they too have a role in reducing state/community conflicts and that their presence in partnership with commercial and governmental organizations produces synergy. This means that far more of value is produced than the sum of the individual parts.

This 'more inclusive' state strategy has been progressively elaborated and has been shaped in the face of criticism of the British urban development corporations (UDCs). These organizations were criticized for concentrating upon raising land values in hitherto stagnant inner city land markets at the expense of building the capacity of the local people [22] and basic infrastructure.[23] A new way had to be found, and so during 1991–1992 the **City Challenge** initiatives were unveiled [24]. Resources were removed from existing urban grants and concentrated in the hands of the City Challenge bid winners. It included £45 million of housing investment; an emphasis upon what can be achieved rather than addressing social needs; monitoring outputs; a stronger emphasis upon multi-agency processes, building local growth coalitions – and new partnerships. Unlike the UDCs, the City Challenge designations would include quite large populations, but the competition between inner city areas for the resources which designation would confer continued the system of patronage developed for the UDCs [25].

The City Challenge approach can allow for involvement of local people socially and economically where this has been identified as important, but the City Challenge teams are expected to achieve targets negotiated with central government. These targets are largely about physical items of the built environment rather than the achievements of local people, and so the 'output measures', by which central government appraises the schemes, rarely reflect social outputs unless they are successfully negotiated by the challenge team. Local people have no clearly stated political rights in this process, which means that the practice of the various City Challenges can be extremely varied. Central government motivation for this approach promotes the notions that:

- local people must be stakeholders (the alternative is alienation and future failure);
- targeted groups must be capable of being connected to the economic mainstream or they will flounder as soon as the projects end.

WORKPIECE 8.3

STUDYING WAYS OF INVOLVING PEOPLE

Select one of the approaches to involvement from the list below, and research its use with reference to your own town or district.

- Planning for Real
- Community Urban Design Assistance Teams
- Community Development Trust
- Decentralization of local government function through Neighbourhood Forum or Area Committee
- Environmental Impact Assessment
- Environment Forum
- Tenant Participation
- Highway Procedure
- Public Local Inquiry into adoption or interpretation of a Development Plan.

EXAMPLE 8.4

BIRMINGHAM FOR PEOPLE

By contrast with the Bishopsgate Goodsyard LET scheme, the LET proposal to redevelop the Bull Ring in Birmingham completely ignored local interests, including that of the market traders. The scheme required a long-term land assembly operation, supported by the City Council. It was launched at a time when the land market was booming and office rents in the Birmingham area were rising to an unprecedented level.

From the developers' point of view it was driven by the economic necessities of completely redeveloping a 1960s retail and markets area on the one site and the prospect of achieving a very large increase in office space at the heart of Britain's second city. The site was 26 acres, for which a development was proposed consisting of one million square feet of retail space, 300 000 square feet of office space, 100 000 square feet of restaurant and leisure space and 3000 parking spaces.

From the city council's point of view, the proposal offered a relatively easy way of upgrading the quality of its retail facilities, which had suffered badly through the loss of three large multiple stores and the rivalry of out-of-town regional shopping centres, particularly the Merry Hill centre near Dudley.

Very considerable economic pressures were mounting to give consent to this development. However, pressures from the other side came into play. Firstly, the market traders ferociously resisted the prospect that their trade would be dramatically disrupted during the development process. Secondly, the markets and shops of the Bull Ring offered goods which represented good value for money to the poorer sections of Birmingham's population; thousands shopped in the area every week. Thirdly, the redevelopment envisaged the replacement of a famous 1960s land mark – the Rotunda, a cylindrical office building of dubious aesthetic value but of tremendous sentimental significance.

A group of enthusiasts made up of environmentalists, urban designers, architects, planners, students, residents of the city centre, members of the congregation of local churches and other local people came together to form Birmingham for People (BfP). They were united by a desire to use the time to restore the vitality of the city centre. In launching a project for 'A People's Plan for the Bull Ring', BfP were able to involve in a detailed way about 100 local people through a series of meetings. Their aim was 'to find a form of development (and conservation) that meets the needs of all sections of the community – women and men, young people and elderly, residents and visitors etc.'

A simple but effective 21-page plan was prepared [26] which talked about 'Learning from the past', 'A strategy for redevelopment', 'Removing through traffic', 'Groundspace for pedestrians', 'A flourishing markets area', 'Shopping for all tastes', 'A place to live and work' and 'Phasing of development'. It included an itemized allocation of floor space by uses. This was completed in early 1989.

The LET scheme was criticized for proposing a development on an alien scale, lacking historical continuity with the Bull Ring location. It was described as 'oversized', creating a massive 'indoor' environment and adding to two other schemes which were not very popular. The development would create massive blockages to pedestrian flows in public access routes; it would privatize large parts of the urban area and the whole city centre would be disrupted while the development went ahead. It was criticized for being too expensive, too commercially risky, containing too much shopping space and too few other uses; it would create too much private car traffic and would be an environmental disaster. Nonetheless the City Planning Committee approved the scheme!

In due course the property boom turned to bust and the grandiose scheme of the developers was shelved. Nearly six years later, in January 1995, the original developers (who had since merged with a Swedish development company and become SPP–LET) came back with another scheme, bearing a remarkable resemblance to the People's Plan!

BfP played a very important role in standing out against the extreme commercial pressures that come into play in a property boom. By demonstrating that public involvement can be the watchword of good urban design it strengthened the hand of those in the city who saw the need for considerable improvement in the quality of the city centre. However, since the new Bull Ring redevelopment proposals have been received by the planning department, no additional consultation is envisaged besides that laid down as a statutory requirement. There are still many important issues which will affect the quality of the final outcome.

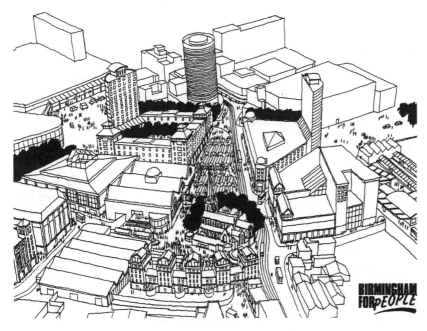

Figure 8.6 Towards a better Bull Ring – a people's plan. (J. Holyoak; reproduced with permission.)

Figure 8.7 Illustrations from the people's plan for the Bull Ring. (J. Holyoak; reproduced with permission.)

WORKPIECE 8.4

VOLUNTARY ORGANIZATION AND COMMUNITY LIAISON

Identify in your own locality a voluntary organization whose objectives include consultation with other local people over matters concerning development. It may be a wildlife group, a building conservation group, a safer environment group or something similar. When you have identified an appropriate organization, obtain copies of any policies or publicity material that explains how they approach others in the community. If possible, seek a brief meeting with the person in the organization most involved in community liaison work.

Compile a small dossier of the approaches used in their community liaison activity.

SUMMARY

Public involvement is appropriate in most cases of development. It is not always very easy to see who should be involved and at what level. The checklist introduced earlier provides a starting point for looking at approaches to involvement.

The European Environmental Bureau has promoted a European Environmental Charter on citizens' rights which highlights the right to participatory democracy including access to information (EU as well as national institutions), access to decision making and access to justice (national courts and the European Court of Justice). The charter espouses the right to environmental quality and the right to human development. It is proposed to include these elements in the revised Treaty on European Union in 1996.

Yet there are clearly very significant differences in the levels of involvement that have occurred for various projects. A comparison is set out in Figure 8.8 for area-based urban regeneration schemes discussed in this chapter. It is arranged in terms of the degree of public participation and the type of state intervention that helped to organize it.

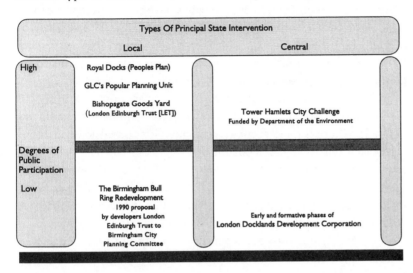

Figure 8.8 Area-based urban regeneration and degrees of community involvement. (Dick Pratt.)

Some of the most challenging difficulties will continue to be those of involving local people in the development of mega-projects. The recent hotly disputed issues are concerned with large-scale infrastructural investment, whether it be high dams for hydroelectricity in Malaysia, wind farms on the western coast of the UK, the Trans-European Road Network (TERN), European fast rail networks and the accompanying redevelopment of railway lands, or the Cardiff Bay Barrage. Such large-scale developments, if approached incrementally, raise the possibility of being accused of realization of policy by stealth.

Locally small-scale developments which make a contribution to the overall common good, but which may be unpopular at the site chosen, will always prove difficult. For example, housing infill development in an area, consent for the establishment of guest houses for patients/offenders discharged from institutions, or the identification of sites for ethnic minority travellers all excite oppositions from local residents.

There may be very many obstacles to effective participation. In the case of private commercial developments these are usually associated with the habitual concern for business secrecy. But could opportunities for gaining wider acceptance for developments be lost as a consequence?

In the case of publicly funded developments, a number of factors may maintain barriers to effective public participation. They include bureaucractic habits and a lack of skills in involving the public. Where commercial or publicly funded organizations have the will and the capacity to involve people, other obstacles can be present. The absence of what might be called 'place solidarity' or 'place affinity' (identification with a place and commitment to it) can mean that efforts made by development agencies prove desultory.

Undoubtedly the way that places and space are currently used will also influence the amount of public involvement. Where a variety of land uses coincide there is more scope than in monofunctional areas. Areas of mixed social groups allow for a pooling of complementary skills so long as differences of interests are not so great that alliances are not possible. Compact settlement (which does not necessarily mean higher density) means that there are more possibilities for arranging social interchanges. Short journeys are good for everything, including participation!

Lastly we should consider whether there are uniform methods to be adopted and standards to be achieved. What constitutes citizen power is viewed differently from one place to another. The interest of minorities should always be safeguarded, and diversity of interests and cultures is something which, given good governance, cities have always provided. Arnstein's ladder continues to remind us of the extent to which tokenistic consultation may be used as a justification for bureaucratic decision making. Even in the same locality, different methods will need to be employed to achieve effective and democratic results. Public involvement is not an exact science.

More experience is being gained by public and private bodies in the art of public involvement. Steps have been identified which can be promoted by public policy. At the same time, voluntary organizations have identified ways in which communities can exert influence from the grass roots. Should everyone concerned with the production of the built environment be judged on the strength of the effectiveness of the community involvement in their last project?

CHECKLIST

● The degree of involvement of people in the development or modification of their environment can only be judged in relative terms.

● The criteria for judging whether involvement and participation is satisfactory will depend upon a variety of cultural, historical, political and psychological factors.

190

- Arnstein's ladder continues to provide an effective set of measures by which involvement and participation can be assessed in a wide variety of different circumstances.
- The price paid for unsatisfactory involvement of local people in development or redevelopment is a loss of potential value for the scheme. In the longer term the price is even higher – alienation and dislocation.
- Effective participation and involvement cannot redeem a poor scheme but will enhance even the best scheme and will certainly mobilize support for adequate schemes.
- The involvement of formally constituted groups is necessary to guard against developers' manipulation of local individuals, but only talking to formal representatives may exclude the possibility of discovering what ordinary individuals may think.
- A variety of procedures for ensuring involvement and participation have been developed by government, consultants and voluntary organizations. Each has its merits and must be selected for its appropriateness.
- Trust can only be established between built environment professionals and local people over a long period of sustained democratic policy implementation. This requires a commitment to be made by government (both national and local), the professions and the voluntary sector.

REFERENCES

1. Crow, G. and Allan, G. (1994) *Community Life: an introduction to local social relations*, Harvester Wheatsheaf, London.
2. Arnstein, S. (1969) A Ladder of Community Participation. *American Institute of Planners Journal*, **35**, 216–224.
3. Department of the Environment (DoE) (1994) *Community Involvement in Planning and Development Processes*, HMSO, London.
4. Skeffington, A. (1969) *Report of the Committee on Public Participation in Planning: People and Planning*, HMSO, London.
5. Loftman, P. and Pratt, D. (1994) Public participation in the UDP process. *Town & Country Planning*, September.
6. Davies, J.G. (1972) *The Evangelistic Bureaucrat: a study of planning exercise in Newcastle upon Tyne*, Tavistock, London.
7. Frankenberg, R. (1969) *Communities in Britain: social life in town and country*, Penguin, Harmondsworth.
8. Gosling, P. (1994) Fair exchange – local communities and the Local Exchange Trading Systems. *Red Pepper*, **6**, November.

9. Stott, M. (1994) How to make work pay without money. *Town & Country Planning*, December, p. 326.

10. Phillips, L. *et al.* (1988) Docklands for the people, in *A Taste of Power: the politics of local economics* (eds M. Macintosh and H. Wainwright), Verso, London.

11. Brownill, S. (1988) The people's plan for the Royal Docks: some contradictions in popular planning. *Planning Practice and Research*, **4**, 15–21.

12. Brownill, S. (1991) *Developing London's Docklands: another great planning disaster*, PCP, London.

13. Brownill, S. (1993) The docklands experience: locality and community in London, in *British Urban Policy and Development Corporations*, (eds R. Imrie and H. Thomas), Paul Chapman Publishing, London.

14. Counsell, G. (1992) Wilderness of frozen assets, in *Independent on Sunday*, 6 December.

15. Fainstein, S.S. (1994) *The City Builders: Property, Politics and Planning in London and New York*, Basil Blackwell, Oxford.

16. Rhodes, C. and Nabi, N. (1992) Brick Lane: a village economy in the shadow of the city?, in *Global Finance and Urban Living: A Study of Metropolitan Change* (eds L. Budd and S. Whimster), Routledge, London.

17. Foreman, C. (1989) *The Battle for Land*, Hilary Shipman, London.

18. Morphet, J. and Pratt, D. (1993) Community planning: some recent UK experience, in *Managing the Metropolis: metropolitan renaissance; new life for old city regions* (eds P. Roberts, T. Strutners and J. Sacks), Avebury, Aldershot.

19. Kline, L. (1994) Talk given as Director of Tower Hamlets City Challenge in September.

20. Ward, C. (1989) *Welcome Thinner City: Urban Survival in the 1990s*, Bedford Square Press, London.

21. Ward, C. (1994) Coin Street – exception or example? *Town and Country Planning*, **63**(6), 166–167.

22. Turok, I. (1992) Property-led urban regeneration: panacea or placebo? *Environmental and Planning A*, **24**(3), 361–379.

23. Imrie, R. and Thomas, H. (eds) (1993) *British Urban Policy and the Urban Development Corporations*, Paul Chapman Publishing, London.

24. De Groot, L. (1992) City Challenge: competing in the urban regeneration game. *Local Economy*, **7**(3), 126–209.

25. Coulson, A. (1993) Urban development corporation, local authorities and patronage in urban policy, in *British Urban Policy and the Urban Development Corporations* (eds R. Imrie and H. Thomas) Paul Chapman Publishing, London.

26. Holyoak, J., Newson, J. and Jackson, D. (1988) *Towards a Better Bull Ring – A People's Plan*, Birmingham for People, Birmingham.

Ambrose, P. (1994) *Urban Process and Power*, Routledge, London.

Blackman, T. (1995) *Urban Policy in Practice*, Routledge, London.

Crosby, T. (1973) *How to Play the Environment Game*, Penguin, Harmondsworth.

Fainstein, S., Gordon, I. and Harloe, M. (eds) (1992) *Divided Cities: New York and London in the Contemporary World*, Blackwell, Oxford (especially Chapters 1 and 2).

Greed, C. (1994) *Women and Planning: Creating Gendered Realities*, Routledge, London.

Gibson, T. (1984) *Counterweight: The Neighbourhood Option*, Town and Country Planning Assocation, and Education for Neighbourhood Change, Nottingham.

Gibson, T. (1986) *Us Plus Them: How to use the experts to get what people really want*, Town and Country Planning Association, Nottingham.

Jacobs, B.D. (1992) *Fractured Cities: Capitalism, Community and Empowerment in Britain and America*, Routledge, London.

Little, J. *et al.* (1988) *Women and Cities: Gender and the Urban Environment*, Macmillan, London.

Twine, F. (1994) *Citizenship and Social Rights: The Interdependence of Self and Society*, Sage, London.

Ward, C. (1990) *The Child in the City*, Bedford Square Press, London.

Young, I.M. (1990) *Justice and the Politics of Difference*, Princeton University Press, Princeton.

RENEWAL AND REGENERATION

COLIN WOOD

THEME

The renewal and regeneration of older towns and cities has had a major influence on the shape, nature and uses of urban areas in many parts of the world. It is the purpose of this chapter to set out the main periods in the renewal and regeneration of older urban areas drawing mainly on examples from the UK and the USA; to explore the forces and motivations underlying the different approaches identified; and to highlight the impact these different approaches have had on people and places.

OBJECTIVES

After reading this chapter you should be able to:

● describe and explain some of the main forces which contribute to change and decline in older urban areas;

● identify and place in their historical context the main phases in the evolution of approaches to dealing with the problems of older urban areas;

● list the different ideological perspectives defining the nature of urban problems and explain how these have influenced intended solutions;

● outline some of the main roles of the built environment professions in influencing urban change;

● assess the implications of different approaches to renewal and regeneration on different social groups and different parts of the city;

● critically reflect upon the strengths and weaknesses of both broad approaches and particular initiatives.

DEFINITIONS OF TERMS USED

A brief discussion is merited on the different terms used to describe the processes which form the basis for consideration in this chapter. Phrases such as 'urban renewal', 'urban redevelopment', 'urban regeneration', etc. are often used interchangeably in everyday speech and, indeed, in many academic texts as well. In this chapter, however, the term **redevelopment** will be used to describe the process of rebuilding following the clearance of previous structures, most evident in the comprehensive housing clearance and redevelopment programmes in the 1950s, 1960s and early 1970s which typified many older towns and cities in Britain and the US. **Renewal** will be used to denote the improvement and rehabilitation of the existing fabric of properties,

perhaps best exemplified by the housing improvement programmes which postdated comprehensive clearance and redevelopment in the UK. **Urban regeneration** has a rather wider connotation than the other two terms. It is popularly used to describe the social and economic, as well as the physical, improvement of older urban areas, and it will be used in this sense here.

Renewal and regeneration, though a topical issue in the 1990s, is not a new phenomenon. Modern attempts to improve housing and working conditions date back to the middle of last century and parallel the industrial growth and development of most of our older manufacturing towns.

This chapter identifies a number of approaches which have been used in dealing with urban problems, discussing them in their chronological order. Initially there were the attempts by enlightened industrialists to improve living and working conditions for their workers. These helped to provide the inspiration for the Garden City movement, with plans for new communities in new surroundings, in a cleaner and fresher environment. In turn, these ideas helped to inform the New Towns programme as part of the reconstruction effort after World War II. Taking place in parallel with the growth of new towns in the 1950s and 1960s were the major slum clearance and redevelopment programmes of older towns and cities on both sides of the Atlantic. These were followed, in the UK in the late 1960s and the 1970s, by programmes to improve and rehabilitate older houses and their environments rather than to replace them. Around the same time, increasing attention was also being focused by central government on perceived social problems in older urban areas, and programmes were initiated to try to combat 'social malaise' as well as unsatisfactory physical conditions.

In the 1970s concern mounted about the decline of manufacturing jobs in basic industries in larger towns and cities in the UK and in the United States, with the 'rust belt' cities of the American north-east and the older industrial towns of the north and midlands of mainland Britain being particularly badly affected. Initiatives sponsored and promoted by central and local governments were increasingly directed towards tack-

INTRODUCTION

195

ling the problems of 'de-industrialization', job losses and rising unemployment. In the UK this concern was registered in the form of legislation which recognized the plight of the older urban areas, introduced new measures for dealing with them, and signalled the end of the New Towns programme.

In the 1980s, in both the UK and the USA, concern about economic decline in urban areas remained high on the political agenda. In Britain, new agencies responsible to central government were given the powers and resources to regenerate older urban areas, and measures were taken to reduce the role of local municipalities and remove the layers of bureaucracy which were seen by government ministers as being largely responsible for thwarting private sector investment initiatives. More recently new measures have been introduced which have placed greater faith on competition between localities to stimulate more innovative ways of promoting urban regeneration; which have encouraged a more comprehensive approach towards dealing with problems in older towns and cities; and which have been geared towards coordinating and streamlining the funding regimes operated by central government. The following sections will consider each of these approaches by briefly assessing the forces that drive them, and their consequences for people and places.

MODEL VILLAGES AND GARDEN CITIES

Amongst the best-known of the early attempts to improve housing and environmental conditions were those by enlightened industrialists in the middle and latter part of the last century. The industrial revolution signalled the rapid growth of urban areas, especially in the north and midlands of the UK and in the central region of Scotland. Cheap, cramped housing was built at high densities to accommodate the workers employed in the new factories and workshops nearby. Poor sanitary conditions bred ill-health, and during the nineteenth century there were numerous outbreaks of cholera and dysentery. However, there emerged during this period a number of industrialists who demonstrated a greater concern for the welfare of their workers than their contemporaries. They provided homes for their employees which enjoyed much better standards of space, privacy and lighting than dwellings typical of the time. Moreover, they were built at lower densities and in settings that provided a richer and more varied environment than the crowded and squalid areas that formed the inner suburbs of expanding industrial towns. These model estates are still in evidence today at places like Port Sunlight on Merseyside, New Earswick near York, Bournville in

Birmingham, New Lanark near Glasgow and Saltaire, Bradford. They also helped to inspire the work of Clarence Stein and Clarence Perry in the United States, and many others throughout the old Commonwealth.

Cottage Homes, Primrose Hill, Port Sunlight.

Figure 9.1 Cottage Homes, Primrose Hill, Port Sunlight: a contemporary postcard. (Courtesy of Dr Richard Turkington.)

It may be that the motives for providing this better quality housing in more attractive surroundings were not always entirely altruistic. The belief that a happy and healthy workforce was also a more productive workforce may well have influenced the thinking of these progressive industrialists. Whatever the reasoning, these estates still provide sound accommodation today and bear eloquent testimony to the vision of their creators. They also helped to influence the thinking of some of the founding fathers of modern town planning, and contributed to the development of the Garden Cities movement instigated by Ebenezer Howard.

Howard's ideas were set out in a seminal book published initially in 1898 [1]. In it, Howard presented his case for new settlements based on community values, fuelled by a spirit of cooperation and sharing. They would be free-standing and separate from existing built-up areas, and would combine the best of both worlds – the liveliness and diversity of towns, and the tranquillity and cleanliness of the countryside. The most famous practical realizations of this philosophy are to be found at Welwyn Garden City and at Letchworth.

Figure 9.2 Letchworth Garden City.

The work of Howard and the other founding fathers provided much of the inspiration for measures in the early part of the twentieth century to improve housing and environmental conditions, with town planning appearing in British legislation for the first time in 1909.

Many of these reforms were underpinned by the belief that improvements to living conditions would also stimulate a feeling of spiritual well-being and social harmony – a belief that has informed planning throughout much of the century.

INTER-WAR PLANNING

In the UK, the period after World War I provided an important era for study. Urban growth and expansion accelerated as the municipalities were encouraged to provide 'homes fit for heroes' through public sector house-building programmes. Housing for owner-occupation also contributed to the trend towards suburbanization, as private housebuilders took advantage of lower construction costs and cheaper land on the periphery of towns, and potential purchasers benefited from the growth of building societies and lengthier mortgage repayment periods. Town planning schemes during the 1930s were largely voluntary and only required for the larger local authorities. They were lengthy to prepare and approve, and were essentially regulatory and restrictive. Legislation was passed in 1935 to limit 'ribbon development' along major roads and prevent the coalescence of adjoining towns. There was growing concern at this time about the outward expansion of larger urban areas, the threat this posed to the countryside and agricultural production, about congestion in the larger towns and cities, and about rising unemployment and population decline in the industrial regions following the Great Depression. These concerns prompted a series of major studies which were to provide the foundation for much of the UK's planning and housing legislation soon after World War II [2]. Some of the issues at stake are applicable in other parts of the world.

World War II left much of the UK housing stock either demolished or severely damaged, and much of the early post-war effort was geared towards eradicating the housing shortage. Patrick Abercrombie produced an ambitious plan for the reconstruction of London which envisaged the large-scale decanting of people to a series of satellite towns beyond a green *cordon sanitaire* to enable redevelopment of older housing areas at lower densities to take place. The New Towns programme which followed owed much to Abercrombie's ideas, and to the vision of Lord Reith, the government minister largely responsible for masterminding it. It was part of a comprehensive package of post-war measures intended to ease congestion in the conurbations, provide new homes, prevent urban sprawl and create focal points for regional economic growth.

The New Towns programme was established in three phases. The first phase in the late 1940s and early 1950s was essentially geared towards easing pressures on London and resulted in the designation of 14 new towns, including Stevenage, Harlow and Crawley. The second phase in the early 1960s was closely linked to broader economic planning policies and the alleviation of problems in distressed areas; it included Peterlee and Washington in north-east England. The third phase focused on larger towns and cities and was as much concerned with mobility, accessibility and transportation planning as with housing. Examples here included Milton Keynes (Figure 9.3) and Telford.

The new towns were run by development corporations which had wide-ranging powers and whose members were appointed by central government. There was much opposition in the early days to these new town development corporations from the existing local authorities who saw their powers being usurped by these new agencies, and by the local population who felt threatened by the influx of newcomers [5].

More than 30 new towns were designated in England, Wales, Scotland and Northern Ireland between the early 1950s and the early 1970s. Although much has been written about the new towns in general and on individual new towns, there has been remarkably little systematic research into the programme as a whole [6]. The first phase of new

POST-WAR RECONSTRUCTION AND THE NEW TOWNS PROGRAMME

towns in particular were based on the twin concepts of 'balance' and 'self-containment'. These terms were never clearly articulated but, in essence, it was the expectation of new town planners that these new settlements would house a representative cross-section of the population as a whole; and would contain the full range of activities necessary for a full and productive life without the need to journey elsewhere.

EXAMPLE 9.1

MILTON KEYNES NEW TOWN

Milton Keynes was designated in 1967 and was initially planned for a population of 250 000. The site incorporated the town of Bletchley, about 80 km from London, together with 13 smaller villages. The flexible master plan was based on the principles of freedom of choice, open-ended design to allow for later development, and personal mobility. The plan was based on a system of kilometre grid squares, with a population of about 2000 in each [3]. The design was heavily influenced by the ideas of American planners Melvin Webber and Christopher Alexander, and sought to exploit a range and variety of social opportunities. By promoting low-density development based on an anticipated 100% car ownership, the master plan challenged the conventional wisdom of earlier new towns that relatively high density, low-rise housing helped to promote a sense of 'community'. Some critics see Milton Keynes as representative of a shift in planning principles, to accommodating 'utilitarian individualistic values' rather than shaping or restricting them [4].

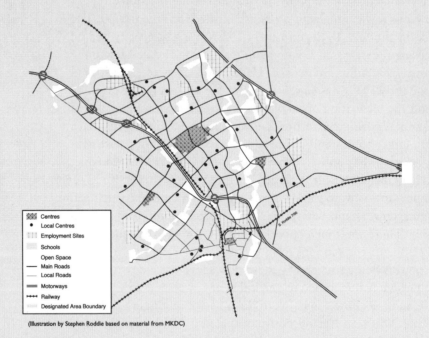

(Illustration by Stephen Roddie based on material from MKDC)

Figure 9.3 Milton Keynes New Town Development Plan showing the flexible framework based on 13 original settlements, allowing for growth, adaption and a high level of personal mobility by private transport.

Many of the early new towns were built on the 'neighbourhood principle' with housing areas planned to a size sufficient to support a range of local services and laid out in such a way that these could be reached within easy walking distance from home. Underlying this approach to planning was the belief that a 'community spirit' could be created by physical propinquity and a development design which facilitated casual social interaction. Some sociologists remained sceptical of this viewpoint, arguing that residents' social class and lifestyle characteristics were likely to have more of an influence on whether they got along together than how closely they lived to, or how often they saw, each other [7,8].

The British new towns are regarded by many as town planning's greatest achievement and the approaches adopted have been applied to new towns throughout the world, including the Middle East and China. Notable examples elsewhere include Chandigarh and Brazilia. At the same time, they have not been without their critics. So what has been said and written about their strengths and limitations?

NEW TOWNS – PROS AND CONS

While it is dangerous to generalize about new towns, given the variety of their locations and circumstances, there is little doubt that they set new standards in terms of quality of design for buildings of many different types, and gave far greater priority than had hitherto been the case to landscaping and environmental treatment. They were managed by people who were committed to the ideals on which they were based and who were responsible for many innovative approaches towards planning and transportation. They sought to achieve a social mix which contrasted with the single-class suburbs of most other towns and cities. Surveys of residents suggested that most people who lived in new towns liked them.

At the same time, and despite the accolades from architects, they were sometimes criticized for being monotonous and spartan in their design. The newer industries and services which set up in the new towns often attracted more skilled and white collar workers, with fewer jobs available to lower income and unskilled personnel. There were reported tensions between newcomers and long-term residents, and claims that some people who had left family and friends for a new life in a new town, housewives in particular, were experiencing 'new town blues'. In many new towns social, leisure and recreational facilities followed a long time after the houses were built, contributing to a feeling of alienation amongst the local population. As national population projec-

201

tions were revised downwards the need for new towns began to be questioned, and growing concern in the 1970s about the economic decline of older towns and cities finally led to the end of the New Towns programme in Britain.

WORKPIECE 9.1

DECENTRALIZATION V. 'URBAN CONTAINMENT'

List the potential advantages and disadvantages of:

1. promoting the decentralization and suburbanization of towns by encouraging development and redevelopment at lower densities, and by creating new settlements;

2. encouraging 'urban containment' through development and redevelopment at higher densities

within existing urban areas and by enforcing strict Green Belt policy.

• Which people tend to benefit most under alternative (1), and why?

• Which people tend to benefit most under alternative (2), and why?

HOUSING REDEVELOPMENT

Renewal of existing towns and cities is as important as new town building. In the UK by the mid 1950s, with the first phase of new towns well under construction and the immediate post-war housing shortages having been substantially overcome, most of the older British towns and cities turned their attention once more to the quality (as distinct from the quantity) of housing. The slum clearance programmes of the 1930s, which had been interrupted by the outbreak of World War II, were revived. What prompted these programmes and what effects did they have?

There were many forces driving the municipal bulldozer, the most clearly articulated of which was the expressed aim of government (both central and local) to eradicate the worst housing inherited from the previous century. There may have been other motivations as well. Many local authorities seized on slum clearance and redevelopment as an opportunity to recast the urban fabric of whole swathes of the city, replacing outworn infrastructure and planning new road systems on the basis of anticipated increases in the use of private transport. Some may have seen slum clearance as a chance to replace old, largely privately rented housing, with more remunerative or prestigious forms of development. In many cities and towns, areas of older housing were replaced by commercial development, or by higher education campuses, for example. Suspicions were also voiced in some quarters that the clearance of privately rented housing and its replacement by new municipal stock was a device used by some authorities to influence the social composition and voting patterns in inner-city wards. Slum clearance and redevelopment became almost a demonstration of civic virility as cities vied with each other to knock down and rebuild the most homes.

The physical environment of large tracts of older towns and cities was transformed (Figures 9.4 and 9.5), as terraces of older privately owned properties along with old factories and workshops were replaced by homes provided by the council, often in the form of flats or maisonettes. Subsidies from central government encouraged the use of innovative building techniques and prefabricated materials to speed the production of dwellings, and the construction of 'system-built' deck access blocks and high-rise flats was widely promoted. Many of these towers and blocks present problems for today's residents, including condensation and damp penetration caused by design faults.

Opposition to slum clearance and comprehensive redevelopment mounted in the 1960s. Where did this opposition come from and what were the arguments? People who had lived as tenants in privately rented housing with the benefit of controlled rents objected to having to pay significantly more, even for admittedly superior accommodation, when they were rehoused by the council. Owner-occupiers objected to the compulsory purchase of their homes as the clearance programme moved into areas of rather sounder housing. There was growing uneasiness about the type of dwellings and form of construction that was replacing the old 'slums'. People decanted to 'overspill' estates on the city outskirts (redevelopment at lower densities usually meant that not everybody could be rehoused in their original neighbourhood) complained about the lack of social facilities and the long journeys back to work or to see friends and family. Their arguments were supported by some academics who held slum clearance responsible for the break-up of communities and saw it as a threat to working-class areas [9].

This opposition coincided with the findings of a national survey into housing conditions which found that the scale of poor-quality housing could not be dealt with by clearance alone [10] and by growing concern about the impact of expensive redevelopment programmes on the public purse at a time when the government was facing balance of payments difficulties. Objections to high-rise housing reached a crescendo when a tower block in London partially collapsed as a result of a gas explosion in the late 1960s. This saw a policy shift as more attention was directed towards the improvement of older, sounder housing in parallel with the continued clearance of the oldest and poorest stock.

OPPOSITION TO CLEARANCE

EXAMPLE 9.2

COMPREHENSIVE CLEARANCE AND REDEVELOPMENT OF HULME, MANCHESTER

Manchester's post-war slum clearance and redevelopment programme spanned some 20 years from the mid 1950s to the mid 1970s, during which period the city acquired and demolished some 88 000 older dwellings. The redevelopment of Hulme, an inner-city area of cramped, back-to-back terraced houses and old workshops just to the south of the city centre, formed an early part of the programme (Figure 9.4). Redevelopment substantially eradicated the old grid-iron pattern of streets and incorporated plans for a number of crescents – deck-access curvilinear system-built blocks of flats and maisonettes intended by their designers to replicate the Georgian terraces of Bath. They were even named after famous architects of the past.

Problems began to emerge only a few years after their completion. Many of the residents complained of condensation and damp penetration; and design faults and poor workmanship in the original construction soon became apparent. The city council was unable to take action against the main contractors, who in the meantime had gone into liquidation. The flats and maisonettes on upper storeys proved to be unpopular, especially for families with small children. Vacancies increased and homes in the Crescents became more difficult to let. For a time the city council offered some accommodation to single-person households and students at the nearby universities, but as maintenance costs escalated and the Crescents developed a reputation for crime and violence, more radical solutions were considered. In the early 1990s Hulme was the subject of a successful bid for funding under the government's urban regeneration City Challenge initiative. The Crescents were demolished and plans approved to rebuild much of the land with conventional low-rise homes.

The design framework for the 'new' Hulme stresses, amongst other things, the importance of the street as a public space to promote socialization; that new homes should be provided at a density which will sustain a wide range of services; that all parts of Hulme should be equally accessible for all people, including women, children, elderly people and physically disabled people; and that the design of Hulme should cater adequately for the car without encouraging its use [11].

Figure 9.4 Comprehensive clearance and redevelopment of Hulme, Manchester.

Figure 9.5 More intimate traditional later redevelopment, Moss Side.

SLUM CLEARANCE

With reference to official reports, newspaper cuttings and other documented evidence, describe the process of slum clearance for part of a town with which you are familiar.

- Why did the local authority choose slum clearance instead of, say, housing improvement or just leaving the area as it was?
- What was the reaction of residents directly affected?
- Were there any differences in reaction between, say, long-term residents and newcomers, or between owner-occupiers and tenants?

HOUSING RENEWAL

The decision to renew or redevelop has important consequences for the character and qualities of the resultant environment. In the late 1970s additional grant assistance encouraged owners and landlords in England and Wales to improve older properties, particularly in specially targeted general improvement areas (GIAs). GIAs were areas of basically sounder older housing where, after appropriate improvements both to properties and to the surrounding environment, the dwellings could be expected to provide adequate accommodation for the next 30 years. Area-based improvement of this nature was seen as a way of preventing the decline and deterioration of older housing; as a way of retaining local 'communities'; and as a less costly solution than comprehensive clearance and redevelopment. As such, the policy had widespread backing and all-party support, and by the early 1970s improvement activity was matching or exceeding clearance and new building [12].

However, it was not long before this new approach to renewal was also coming under scrutiny. Studies showed that in some areas landlords were capitalizing on the more liberal grant regime by using public funds to improve their properties and then re-letting them at much higher rents or selling them for owner-occupation at considerable profit. Stories were reported of 'winkling' (landlords offering tenants financial inducements to move out of their homes so they could be improved, re-let or sold) and physical harassment to achieve the same ends. It was also claimed that rather better-off owner-occupiers were also benefiting as they were more able to meet the balance of total improvement costs not covered by the government grant. In some more fashionable parts of larger towns and cities, and in London particularly, the social character of neighbourhoods was changing as more affluent newcomers purchased improved homes that locals could no longer afford: a process known as 'gentrification'. Ironically, a policy that had been introduced in part to safeguard local communities was having the opposite effect.

Further legislative changes in the UK, introduced in the mid 1970s sought to target properties and people in greatest need and to tackle the problems of 'grant abuse'. A policy of 'gradual renewal' also signalled the end of the era of wholesale clearance and new council building [13].

More recently, renewal areas (RAs) have replaced previous area improvement initiatives for dealing with concentrations of older private sector housing. These are generally larger than the schemes which they replace. They are intended to combine clearance and new build activity with improvement where appropriate and aim to attract significant private sector investment. RAs are identified following a comprehensive neighbourhood assessment which takes account of physical and social conditions and the costs of different renewal options. Programmes are implemented within the context of a 10-year strategy [14].

WORKPIECE 9.3

HOUSING IMPROVEMENT

Which of the following groups of residents are likely to be the main beneficiaries of housing improvement as distinct from slum clearance and redevelopment, and why?

- owner-occupiers;
- landlords of privately rented homes;
- tenants of unfurnished privately rented accommodation;
- tenants of furnished privately rented accommodation;
- lodgers;
- long-term residents;
- newcomers.

From the list above, which groups of residents are likely to lose most from housing improvement as distinct from slum clearance and redevelopment, and why?

THE 'URBAN MANAGERS'

The rapid changes to the physical fabric of, and in some cases the social relationships in, many parts of older towns and cities led to increasing attention being focused on the people and professions seen to be largely responsible for these changes. This interest was primarily initiated by a study of the workings of the housing market and housing allocation mechanisms which focused on landlord/tenant relationships in a 'twilight' area, and on the role of private landlords who provided accommodation for those who could not afford owner-occupied dwellings and who did not qualify for local authority housing [15]. Other studies investigated the influence of estate agents in limiting the access to certain neighbourhoods of potential buyers deemed to be unsuitable by employing discriminatory screening techniques [16]; and the lending policies of some building societies which refused funding for house purchase or improvement to applicants in 'red-lined' inner city districts [17].

The growing body of research into 'urban managerialism' increasingly concentrated on the roles, actions and motives of urban managers in the state sector, at a time when public policy and government spending was having a significant impact at the local level. A wealth of literature accumulated on the role of local authority housing officers and their housing allocation policies in influencing access to housing [18]; the difference in perceptions and values between professionals and residents over the issue of 'unfit' housing [19]; and the distributional consequences for people and places of land use policies devised and implemented by town planners [20]. Addressing the key questions of 'who gets what, when, why and how?' became fundamental matters of concern.

The increasing awareness of the effects of physical change on social relationships and the greater emphasis on housing improvement and neighbourhood renewal in the UK coincided with broader shifts in urban policy. The 're-discovery' of poverty despite the best efforts of the Welfare State, the growing problem of homelessness in many major cities, public concern about immigration to Britain from the New Commonwealth and the experience of various urban initiatives in the United States all helped to capture the attention of politicians and leading opinion formers in the UK.

Early attempts to tackle the 'urban problem' concentrated on educational initiatives. It was the belief of government and urban policy-makers that better schooling during the formative years would enable youngsters to break free from the cycle of deprivation into which their families and forebears had been trapped. The Urban Programme, established in 1968, sought to improve educational and social conditions in tightly drawn targeted areas defined on the basis of census indicators, used for the first time for this purpose. This 'spatial targeting' has remained a key feature of government urban policy ever since but it has been criticized on a number of counts. Firstly, not everyone within these areas may be 'in need'. Secondly, there are many more people in need outside these defined areas than there are within them. Thirdly, preferential treatment for people or properties inside these 'special areas' overlooks or ignores the very similar situations that are often to be found just outside them. It could be argued that targeting people is a more effective way of tackling poverty and deprivation than targeting places.

THE BEGINNINGS OF URBAN POLICY

THE URBAN PROGRAMME AND SPATIAL TARGETING

WORKPIECE 9.4

SPATIAL TARGETING

What are the main advantages of targeting defined geographical areas for special treatment? What are the main drawbacks?

COMMUNITY DEVELOPMENT PROJECTS

The declining manufacturing base in many older industrial towns and cities, and the increasing levels of unemployment that this created, has added to problems of urban deprivation and heavily influenced a further series of studies and projects which shifted the focus of urban policy more towards economic issues. The most radical of these initiatives in the UK were the Community Development Projects (CDPs). They were set up initially as action–research projects to investigate and make recommendations to deal with multiple deprivation in specific areas. However, they soon came to question the conventional wisdom that urban problems occurred as a result of socially transmitted deviant pathologies. Rather, they argued, urban deprivation resulted largely from poverty created by wider economic forces, and neither the causes nor the solutions could be found within narrowly defined geographical boundaries. The CDP conclusions in many ways reflected the spirit of the times. They argued for radical solutions to address the underlying structural causes of inequality, which meant, in effect, a shift in the balance of political power in favour of the 'have-nots' in society. The CDP message was clearly uncomfortable for the Establishment and the projects were soon wound up, but they helped to promote a fundamental re-think of the way urban policy was to operate [21].

INNER AREA STUDIES AND POLICY

Amongst the most influential of the various studies of the 1970s were the UK inner area studies which were undertaken by private consultants. These studies, carried out in Lambeth, Liverpool and Birmingham, accepted that urban problems of population decline, economic disinvestment and social polarization were too complex to be tackled on a piecemeal basis by individual agencies. Instead, they advocated a 'total approach' which would require a comprehensive and coordinated attack on urban problems by all of the organizations and agencies responsible for service funding and delivery in depressed areas, and with the full involvement of the residents of those areas. They also realized that additional public funding would be needed to deal with bad housing, unemployment and low incomes. The lessons learned in these studies have influenced thinking in many countries, including the fast-growing cities of the Pacific rim.

The inner area studies paved the way for legislation to tackle urban problems and marked the end of the New Towns programme. This Inner Urban Areas Act, 1978 represented probably the most comprehensive package of measures specifically geared towards addressing the problems of older towns and cities in the UK ever enacted by central government. For some time considerable disquiet had been expressed by town and city councils in the major conurbations about the diversion of scarce public funds towards the provision of infrastructure to attract jobs and investment in the new towns. Now, instead, the larger, older towns and cities were to benefit from increased funding. The Urban Programme was to be expanded, with partnership arrangements established to administer the extra funds in the inner areas of the major conurbations and 'programme' status conferred on other larger towns with serious urban problems. In addition, central government resources to local government were to be 'bent' to favour older urban areas, and mainstream spending programmes for services such as housing and education were to be enhanced. Urban partnership and programme areas were given special powers to designate industrial and commercial improvement areas (IIAs and CIAs) in an attempt to secure jobs and investment. In all of this, the local authorities were seen as 'the natural agencies to tackle inner area problems' with their wide powers, resources and local democratic accountability [22].

During the 1980s, in the USA and UK, the perception of the causes of urban blight differed from those of earlier years and the methods of tackling it differed accordingly. In the 1960s and 1970s, urban decline had been variously perceived to arise from a bad physical environment, a 'culture of poverty', poor urban management and use of resources, or structural inadequacies. In the 1980s, it was the 'dead hand' of state bureaucracy (and local state bureaucracy in particular) that was seen to be one of the main causes of decline. It was argued that the local state apparatus was thwarting private investment and entrepreneurial activity, and steps were taken on both sides of the Atlantic to 'free up the market'. The measures for achieving urban regeneration – now perceived very much in terms of creating development and employment opportunities – stressed the importance of liberalization and enterprise and the role of the private sector, as well as re-emphasizing the need for better coordination within and between agencies.

During the 1980s a vast array of initiatives were introduced by central government in the UK to tackle the problems of declining urban

URBAN REGENERATION

areas. Many of these sought to bring derelict land or buildings back into use, to target resources more selectively, to encourage private investment (often on the back of public pump-priming), to coordinate the activities of agencies operating in declining areas and to tackle the problems of larger, poorer local authority housing estates by promoting the 'right to buy', tenant participation and tenure diversification. Two of the flagship initiatives were the enterprise zones (EZs) and the urban development corporations (UDCs).

ENTERPRISE ZONES

Enterprise zones were established in 1980 and in many ways best exemplify central government's preoccupation at this time with deregulation. Based on ideas imported from Hong Kong and other cities in the Far East, they were seen as areas where the 'free market' would be allowed to operate with the minimum of state controls and restrictions in order to encourage entrepreneurial activity and private investment. Schemes for EZs were to be drawn up locally at the invitation of the Secretary of State who would make the final designation. Firms within EZs were to enjoy a number of financial and administrative advantages, including freedom from local taxes for a 10-year period (the local authorities were to be compensated by the Treasury), tax allowances for capital expenditure on buildings, and a simplified planning regime which granted automatic permission for approved types of development. Enterprise zones sat uneasily at the interface of urban and regional policy. While some designations were made in areas of urban dereliction and deprivation, others were made to promote investment in depressed regions in locations divorced from typical inner-city problems.

Enterprise zones have been much researched and government support for them appeared to wane towards the end of the 1980s when it was revealed that perhaps fewer than 14 000 net new jobs had been created at a cost of between £23 000 and £30 000 each. Other research findings indicated that EZs created a dual land market, with values higher within the zone than in immediately adjoining areas; that designation and the financial advantages this conferred helped to subsidize marginal firms rather than encourage innovation; and that the main beneficiaries were likely to be investors and landowners rather than those directly involved in production. Early concerns that the relaxed planning regime would lead to a fall in design and development standards were not supported by research published in the mid 1980s [23].

210

UDCs were introduced in the same legislation which established EZs in 1980. The first two corporations were established for the London Docklands area (LDDC) and for Merseyside in 1981, and further rounds of designations in early and late 1987 included Trafford Park in Greater Manchester, the Black Country in the West Midlands and Cardiff Bay, with 'mini' UDCs for central Manchester, Leeds and Bristol. More recently UDCs have been declared for Plymouth and for the Heartlands area of Birmingham. UDCs are run by boards whose members are appointed by the Secretary of State. They normally include people with a good working knowledge of the urban development area for which the UDC is responsible, and people with close links with the business community. The development corporations are funded directly from central government, but can also raise finance through the sale of assets (land and property) and by borrowing. They have a wide range of powers to acquire, reclaim and dispose of land, and provide for basic infrastructure, with a general remit to:

> secure the regeneration of [their] area by bringing land and buildings into effective use, encouraging the development of existing and new industry and commerce, creating an attractive environment and ensuring that housing and social facilities are available to encourage people to live and work in the area.[24]

In many respects, UDCs have carried forward the spirit of the new towns corporations, but instead of creating new settlements in predominantly rural surroundings, their job has been to tackle dereliction and loss of confidence in existing urban areas. Much has been written about UDCs, so what are their strengths and what are their drawbacks?

UDCS – PROS AND CONS Urban development corporations claim a number of advantages in promoting urban regeneration in their areas. They are single-minded organizations with a sole duty to achieve the physical transformation and economic upgrading of their areas, and they are unencumbered by the many other responsibilities which, it is argued, make the task for local authorities so much more difficult. They have wide-ranging powers and substantial funding in order for them carry out their remit. They are likely to be able to attract considerable additional finance from the private sector because of board members' commercial contacts and as a result of public sector pump-priming to reclaim land for development (indeed, the amount of extra private resources 'levered in' is one of the main criteria against which their suc-

URBAN DEVELOPMENT CORPORATIONS

211

cess is judged). Because they have a limited life (up to 15 years for the LDDC and Merseyside UDC; 5–7 years for the 'mini' UDCs) it is expected that this will concentrate their energies and they will show early returns in the form of new development and inward investment. Evidence in the annual reports of the UDCs, and 'on the ground', shows that substantial progress has been made in most areas measured in terms of land reclaimed, homes built, roads constructed and environment improved.

UDCs, especially the LDDC, have also come in for some fierce criticism. The fact that the board members are appointed rather than elected has led to accusations that UDCs are not democratically accountable at the local level and do not always adequately represent the interests of people living in the development area. The LDDC in particular has been accused of using large amounts of public money to fund jobs and homes which have gone predominantly to commuters and 'outsiders' rather than long-term residents of the Docklands [25]. It has been suggested that the 'quick-fix' project-oriented approach of UDCs has neglected longer-term strategic considerations. Much concern has been expressed, for example, about the effects of the UDC-backed marina development in Cardiff Bay on wildlife habitats and adjacent residential areas [26]. Bristol City Council objected to the UDC in central Bristol on the grounds that the area chosen displayed few of the criteria used to identify suitable development areas [27]. The feeling amongst many local authorities is that, given the same levels of funding, they could do the job just as well. Indeed, some UDCs have been criticized for taking the credit for some schemes already substantially completed by the local authorities and other agencies in whose areas they operate [28].

BELIEF IN 'TRICKLE-DOWN'

The taste for flagship projects, much favoured by the UDCs, was also a feature of many municipalities in both the United States and the UK in the 1980s. Much has been written about the success of physically transforming parts of older industrial cities on both sides of the Atlantic, with Baltimore's waterfront redevelopment figuring particularly prominently in the literature. High-profile schemes to attract private investment and development to downtown areas, as opposed to directly targeting improvements at the poorer residential areas, have been justified on the grounds that these schemes will create jobs, encourage spending, boost the local economy and create wealth which will ultimately 'trickle-down' to the depressed neighbourhoods [30].

EXAMPLE 9.3

BIRMINGHAM HEARTLANDS URBAN DEVELOPMENT CORPORATION

Birmingham Heartlands UDC started life in the late 1980s as a partnership venture between the city council and several major private construction firms. At a time when ministers were actively considering UDC designation for this part of east central Birmingham, a new agency was established to demonstrate that local initiative could bring about the effective regeneration of an area suffering from economic decline, a poor environment and limited housing choice without the need for control from central government. A development strategy was approved which was intended to improve the existing housing stock; provide new homes for sale; create new commercial opportunities along the canalside; improve internal transport links; provide the focus for a major new development on a reclaimed site; and effect improvements to existing industrial sites in the area. The strategy envisaged the creation of 20 000 new jobs and £1300 million of investment by the private sector. It was also a central part of the strategy that improvement, development and new investment should benefit existing residents of Heartlands, instead of 'outsiders' as had been the case in some of the UDCs.

While some early progress was made, the recession of the early 1990s meant that private investment was less than hoped for, and because Heartlands was not a UDC it did not benefit from direct funding from central government. In 1992, after discussions between leading local politicians and government ministers, an Urban Development Corporation was established to take over the work of the Heartlands agency. Additional central government funding was committed to the area, and the new board (including a stronger representation of local councillors than UDCs hitherto) was declared in order to meet the city council's concern about the need to maintain local democratic accountability [295].

Figure 9.6 Bordesley Village in Birmingham Heartlands. New affordable owner-occupied and social-rented housing close to Birmingham City Football Club (floodlight in the background) is perhaps the largest 'urban village' development of its kind in the UK.

However, some concern has been expressed about who pays and who gains from this type of investment, especially when tax inducements have been made to attract developments used largely by higher income groups at the expense of funds for local services. Similar concerns have been expressed in Britain where major prestige projects such as the International Convention Centre and the National Indoor Arena in Birmingham (key elements in the City Council's promotion of 'business tourism') have been argued by some critics to have been subsidized in effect by diverting funds from housing and education services. Doubts have also been expressed about the number and type of jobs created by these projects. Some research indicates that relatively few permanent full-time jobs are generated, and many of these are poorly paid [31].

EXAMPLE 9.5

URBAN REGENERATION IN CLEVELAND, OHIO

Cleveland lies on the shores of Lake Erie in the north of the state of Ohio, and forms part of the so-called 'rust belt' of north-east United States – an area which has witnessed massive industrial decline as a result of the collapse of basic manufacturing industries. Between 1970 and 1985 the Cleveland Metropolitan area lost 86 000 manufacturing jobs, and a further loss of 50 000 is projected up to the year 2005.

Population loss, primarily as a result of migration to the suburbs, has accompanied economic decline. The city's population in 1950 was 914 000 compared with an estimated 500 500 in 1990. The forecast for 2005 is 321 800. The city constitutes one of the most racially segregated housing markets in the United States, with the vast majority of Cleveland's black American population living to the east of the Cuyahoga River, and most of the whites to the west. Some 80 000 defective dwellings have been cleared in the city since 1960, but over one third of the housing stock remains substandard. Neighbourhoods such as Hough to the east of the city centre are amongst the poorest in the country. The area has suffered from 'white flight' to the suburbs over the past 40 years, the wholesale abandonment of properties, multi-occupation of older poorer dwellings, poor quality public sector housing and high vacancy rates. In 1978 Cleveland was the first American city to default on Federal loans, registering debts of $111 million. It became the butt of comedy shows, with references to the city as 'the mistake on the Lake' [32].

During the 1970s the main thrust of public policy in Cleveland was aimed at tackling directly the problems of the poorer neighbourhoods. Under planning director Norman Krumholz the idea of equity planning was put into practice. This meant explicitly attempting to influence public and private development and investment decisions across a range of initiatives in a way which would best benefit the residents of Clevelands most deprived neighbourhoods. This approach was based on the belief that private investment decisions, if left to their own devices, would tend to favour the haves rather than the have-nots; and that traditional planning methods concentrating on land-use could make only a marginal impression on the problems of people in the worst parts of the city [33].

Following changes in political control and administration in the 1980s the city council has adopted the approach favoured by many cities in the American north-east. Emphasis has been placed on trying to attract private development and investment into the downtown area in the belief and hope that demand for services will be boosted, spending will increase, jobs will be created, and the benefits will 'trickle-down' to the neighbourhoods. A new convention centre has been built; older warehouses down by the river have been converted into bars and restaurants; there are plans for a new domed stadium for major sporting events; and the city's Rock and Roll Hall of Fame is now open. Nevertheless, there is considerable unease in some quarters that this focus on the city centre is at the expense of the poorer neighbourhoods, and that the tax concessions offered to induce private capital to invest in the city have reduced the amount of public funding available for essential services in deprived areas.

Figure 9.7 Neighbourhood abandonment in Cleveland, Ohio.

Figure 9.8 Downtown Cleveland, Ohio.

More recent urban regeneration initiatives in the UK have stressed the importance of competition for funding between municipalities, ostensibly to encourage innovative approaches, greater coordination by central government in processing and allocating financial resources, and the establishment of a new national agency to promote comprehensive approaches to regeneration at the local level. Critics claim that these are devices to reduce still further the amount of public funding for renewal and regeneration, and may mean that such sums as are available will go to those areas which mount the glossiest presentations rather than to those in greatest need.

MORE RECENT APPROACHES TO REGENERATION

WORKPIECE 9.5

URBAN REGENERATION IN THE 1980s

What have been the key themes of urban regeneration in the 1980s?

In what ways do approaches to regeneration in the 1980s differ from earlier approaches?

Choose one of the initiatives introduced in the 1980s to deal with problems in older urban areas with which you are familiar, and set out clearly what it set out to achieve.

Taking the same example, assess how successful this initiative has been and what its strengths and weaknesses are.

SUMMARY

This chapter has traced the evolution of renewal and regeneration, with particular reference to experiences in the UK and United States. In the mid and late nineteenth century some of the most notable attempts to create better housing and environmental conditions were promoted by enlightened industrialists, and found their expression in model villages like Bournville and Port Sunlight. In the early part of the twentieth century the Garden City movement was developed by utopian visionaries like Ebenezer Howard who experimented with new settlements intended to combine the best of both urban and rural environments, and based on the principles of shared effort and reward.

These ideas helped to inspire the post-war reconstruction of towns and cities in the UK, with the New Towns programme forming a key element in central government's planned decentralization of people and jobs from the congested conurbations.

A period of comprehensive clearance and redevelopment in the 1960s was replaced by one of gradual renewal of older housing. The greater emphasis given in housing programmes to social considerations and public involvement in renewal decisions at this time was mirrored to some extent by early urban policy initiatives. These stressed the importance of better education provision and social service delivery in specially targeted deprived areas.

Concern about population loss and economic decline in older industrial areas led to greater attention being placed on the need to stem decentralization, safeguard jobs and deal with the growing social polarization between the inner cities and the more affluent suburban areas. Urban regeneration initiatives have more recently reflected the belief that the local state was partly the cause of urban decline and that solutions lay in encouraging private enterprise and entrepreneurialism, removing unnecessary bureaucracy and promoting new development.

Figure 9.9 shows how these different approaches to renewal and regeneration reflect different perceptions of the nature of the problem. These perceptions themselves are partly a product of their time but they also reveal the influence of ideology in defining the nature of the urban problem and devising urban policy. Key questions which remain for future renewal and regeneration policy are:

● Who are we trying to help?
● Can any one approach provide the answer?

PERSPECTIVE	PERCEIVED PROBLEM	GOAL	MEANS
Physical determinism	Poor housing and environment	Urban redevelopment	Slum clearance and redevelopment
Culture of poverty	Pathology of deviant groups	Better social adjustment	Education
Cycle of deprivation	Individual inadequacy	Better families	Social work
Institutional malfunction	Planning failure	Rational planning	Corporate planning
Resource maldistribution	Inequitable distribution	Reallocation of resources	Positive discrimination
Structural conflict	Class divisions	Redistribution of power	Political change
The 'New Right'	Local state bureaucracy	Market freedom	Deregulation

Figure 9.9 Different perspectives on the causes of and solutions to the urban problem. (Modified after Coventry CDP, 1975.)

● Is the continuing faith in targeting places rather than people justified?

The experiences of the UK and the United States provide valuable insights and the lessons can be applied in many other contexts around the world.

CHECKLIST

● Terms such as 'urban renewal', 'urban redevelopment' and 'urban regeneration' are often used interchangeably, but they can have different and distinct meanings.
● While it is dangerous to generalize, approaches to dealing with the problems of older urban areas have tended to shift over time, from an emphasis on physical improvements to older areas of housing (firstly by redevelopment, then by renewal), through attempts to improve social conditions, to measures for dealing with economic decline.
● These approaches in part reflect the circumstances prevailing at the time, but also different ideological perspectives on the nature and causes of urban problems.
● The built environment professions have played, and do play, an important role in the renewal of cities. Their importance in rela-

tion to each other may depend on the nature and scale of renewal and regeneration, and the stage in the process.

- The relationship between the central state, the local state and the private sector in dealing with renewal and regeneration have important implications for the process and the product.

- Approaches to renewing older towns and cities take different physical forms, including model villages within existing urban areas, garden cities and new towns outside them, estates of houses built by local authorities in the inner cities and on 'overspill' sites, areas of improved older housing, and high-tech commercial development on derelict and vacant urban sites.

- Different approaches to renewal and regeneration have different social and economic consequences for individuals and groups. These need to be properly understood in order to devise and implement appropriate policy.

- The targeting of discrete geographical areas has been a consistent feature of urban renewal and regeneration policy, but there is a debate about whether places should be targeted for preferential treatment, or people.

- There is also a debate about whether renewal and regeneration should seek to revive the economies of city centres (in the belief that benefits will 'trickle-down' to the residents of poorer areas), or whether emphasis should be given to the direct improvement of conditions and services in the deprived neighbourhoods themselves.

REFERENCES

1. Howard, E. (1898) *To-morrow: A Peaceful Path to Real Reform*, Swan Sonnenschein, London.
2. The **Barlow Report** investigated the distribution of the industrial population; the **Uthwatt Committee** reported on land values, compensation and betterment; and the **Scott Committee** advised on land uses in rural areas.
3. Milton Keynes Development Corporation (1970) *The Plan for Milton Keynes*, MKDC, Milton Keynes.
4. Creese, W.C. (1992), *The Search for Environment: The Garden City Before and After*, John Hopkins University, Baltimore, Maryland.

5. Cullingworth, J.B. (1982) *Town and Country Planning in Britain*, 8th edn, Allen and Unwin, London.

6. Aldridge, M. (1979) *The British New Towns*, London.

7. Broady, M. (1968) *Planning for People: Essays on the Social Context of Planning*, The Bedford Square Press of the National Council of Social Service, London.

8. Gans, H.J. (1968) *People and Plans: Essays on Urban Problems and Solutions*, Basic Books, New York.

9. Vereker, C., Mays, J.B., Gittus, E. and Broady, M. (1961) *Urban Redevelopment and Social Change: a study of social conditions in central Liverpool 1955–56*, University Press, Liverpool.

10. Ministry of Housing and Local Government (1969) *House Condition Survey, England and Wales, 1967*, HMSO, London.

11. Hulme Regeneration Ltd (1994) *Rebuilding the City: A Guide to Development in Hulme*, R. Tym and Partners for Hulme City Challenge, Manchester, June.

12. Roberts, J.T. (1976) *General Improvement Areas*, Saxon House, Farnborough.

13. Balchin, P.N. (1995) *Housing Policy: An Introduction*, Routledge, London.

14. Local Government and Housing Act, 1989.

15. Rex, J. and Moore R. (1967) *Race, Community and Conflict: a study of Sparkbrook*, Oxford University Press, London, for the Institute of Race Relations.

16. Burney, E. (1967) *Housing on Trial: a study of immigrants and local government*, Oxford University Press, London, for the Institute of Race Relations; and Hatch, J.C.S. (1973) *Estate Agents as Urban Gatekeepers*, paper to the Urban Sociology Group of the British Sociology Association.

17. Weir, S. (1976) Red Line Districts, *Roof*, **1**(4), July, pp. 109–114; and Williams, P. (1977) *Building Societies and the Inner City*, Working Paper 54, Centre for Urban and Regional Studies, Birmingham University.

18. Pahl, R. (1970) *Patterns of Urban Life*, Longman, London.

19. Dennis, N. (1970) *People and Planning*, Faber, London; and Dennis, N. (1972) *Public Participation and Planners' Blight*, Faber, London.

20. Eversley, D.E.C. (1973) *The Planner in Society*, Faber, London.

21. Community Development Project (CDP) (1977) *Gilding the Ghetto*, CDP Inter-Project Team, London.

22. Lawless, P. (1989) *Britain's Inner Cities*, Paul Chapman Publishing, London.

23. Cambridge, P.A., Economic Consultants (1987) *An Evaluation of the Enterprise Zone Experiment*, DoE Inner Cities Research Programme, HMSO, London; and Tym, R. and Partners (1984) *Monitoring Enterprise Zones*, Year 3 Report, for the Department of the Environment.

24. Local Government, Planning and Land Act, 1980, Section 136.

25. Brownill, S. (1990), *Developing London's Docklands: Another Great Planning Disaster?*, Paul Chapman Publishing, London.

26. Brookes, J. (1989) Cardiff Bay renewal strategy – another hole in the democratic system. *The Planner*, **75**(1), January, 38–40.

27. Parker, G. and Oatley, N. (1989) The case against the proposed Urban Development Corporation for Bristol. *The Planner*, **75**(1), January, 32–35.

28. Imrie, R. and Thomas, H. (1993) *British Urban Policy and the Urban Development Corporations*, Paul Chapman Publishing, London (for a general review of UDCs).

29. Barnekov, T., Boyle, R. and Rich, D. (1989) *Privatism and Urban Policy in Britain and the United States*, Oxford University Press, Oxford.

30. Wood, C. (1994) Local urban regeneration initiatives: Birmingham Heartlands. *Cities*, **11**(1), February.

31. Loftman, P. and Nevin, B. (1992) *Urban Regeneration and Social Equity – a case study of Birmingham 1986–1992*, Faculty of the Built Environment, University of Central England, Birmingham.

32. Rose, E.A. (1986) Birmingham, UK, and Cleveland, Ohio: An Anglo-American Comparison, in *New Roles for Old Cities* (ed. Edgar A. Rose), Gower, Aldershot.

33. Krumholz, N. and Forester, J. (1990) *Making Equity Planning Work: leadership in the public sector*, Temple University Press, Philadelphia.

FURTHER READING

Altermann, R. and Cars, G. (eds) (1991) *Neighbourhood Regeneration: An International Evaluation*, Mansell Publishing, London.

Berry, J. *et al.* (eds) (1993) *Urban Regeneration: Property Investment and Development*, E & FN Spon, London.

Bianchini, F. and Parkinson, M. (1993) *Cultural Policy and Urban Regeneration: the West European experience*, University Press, Manchester.

Deakin, N. and Edwards, J. (1993) *The Enterprise Culture and the Inner City*, Routledge, London.

Donnison, D. and Middleton, A. (eds) (1987) *Regenerating the Inner City: Glasgow's Experience*, Routledge and Kegan Paul, London.

Gibson, M.S. and Langstaff, M.J. (1982) *An Introduction to Urban Renewal*, Hutchinson, London.

Healey, P., Usher, D., Davoudi, S., Tavsanoglu, S. and O'Toole, M. (eds) (1992) *Rebuilding the City: Property-led Urban Regeneration*, E & FN Spon, London.

Robson, B. (1988) *Those Inner Cities: Reconciling the Social and Economic Aims of Urban Policy*, Clarendon Press, Oxford.

Smythe, H. (1993) *Marketing the City – the Role of Flagship Developments in Urban Regeneration*, E & FN Spon, London.

MAKING CONNECTIONS

DAVID CHAPMAN

THEME

The earlier chapters of this book have explored some of the varied and complex forces which shape the built environment. They have also asked questions about the qualities which we as individuals and communities need for a satisfying life. This chapter looks for connections between the many forces at work, how they relate to diverse communities of interest, and how they may interrelate across different areas of knowledge. How can we interpret what is going on in different situations? How can we recognize the ways of influencing decision making and promoting positive outcomes from corporate and individual actions? This chapter will help to make links between some of the key considerations raised in the previous chapters.

OBJECTIVES

After reading this chapter you should be able to:

● develop a synoptic view of the origins, qualities and uses of the built environment;

● understand the interrelationships between the different actors and interests in the built environment;

● recognize the links between social, economic and development processes;

● consider ways in which professionals from built environment disciplines can enable and facilitate involvement, participation and action;

● understand the roles of public policy and private initiative at both strategic and tactical levels;

- consider the possible impacts of new technologies and communication systems on patterns of living and the future form of the built environment;

- appreciate the importance of creativity, innovation and problem solving in building neighbourhoods and places.

Recognizing the great complexity of the processes of urban change is an important precursor of intervention. How will our human actions affect the shape of the places we live in and the quality of our lives? It is not possible for anyone to understand all of the connections between the forces at work, or the outcomes which may result from them, but we can all consider the possible consequences of our own actions. The ability to make connections between diverse forces and factors not only helps us to understand the competing interests at stake in society but also to seek strategies which respond to and respect those interests now and in the future.

This chapter could not seek to make all of the possible connections which may exist – that would be impossible. What it does is to ask key questions about the broad ways in which we can each reflect on different situations to make connections for ourselves. The themes explored throughout the book, though quite simple in their own terms, present much greater complexity when they are considered together and interrelated. When human aspiration and ambition are added there can be great conflict as well as great collaboration.

The structure of this book has been deliberately conceived to relate three aspects of the processes and products of the built environment:

- The ways in which places develop or decline and what we can learn from the experience of different places and different times.
- The range of physical, emotional, cultural and physiological experiences which places can provide or evoke and what we can learn about the qualities of places. Which tend to be favourable and why do some places fail?
- The ways of shaping or steering change. Who should be involved in deciding and how can participation and decision making be handled?

How can these aspects of our knowledge and experience help us to evaluate each new situation; whether it is a matter of policy, design or management of the built environment?

FORCES, FORMS AND USES

In each of the aspects described above there are factors which we can consider or analyse to try to understand what actually happened or might happen at a particular place or time.

- It is suggested that there are forces at work which influence both the processes and products of the built environment. In each situation we can try to evaluate what the principal forces are and how they may constrain or facilitate action or intervention.
- Built environments take many and various forms. They can offer great enjoyment and great disappointment. What are the factors which lead to one rather than the other? What forms are associated with each and why do we react differently to different places?
- Our reactions to places affect the way we use them, individually and with others. The ways places are used also affects our reactions to them. This interconnection between the form and use of places and its effect upon the resultant experience is an important consideration.

BUILDING NEIGHBOURHOODS AND PLACES

Our principal theme and title is concerned with the ways places are created and occupied, and the opportunities and qualities they offer their occupants. The concept of neighbourhood can be considered as simplistic and romantic; it might be claimed that there are really no such things as cohesive, self-contained neighbourhoods. The idea of neighbourhoods in planning has been very influential, and owes much to the work of Clarence Perry and Clarence Stein in the United States of the 1920s and 1930s [1]. Their plans for integrated mixed use areas where most facilities are within walking distance are being echoed today, with ideas of compact cities and urban villages. But the idea of neighbourhoods as entities which can be created as part of a process of planning and development is not solely what we mean here. Rather we are concerned with the qualities of life and environment which the term 'neighbourhoods' evokes. In practice, the range over which we as individuals seek to satisfy our needs may vary from those of our close neighbours, but some places may satisfy that diversity of need and aspiration more readily than other places. Why is this so? What can be done to improve things? Who can initiate or implement change, and how? These are the sort of questions we wish to raise. There are no standard answers, but by understanding some of the issues at stake we can all make a well-informed contribution.

CONSIDERING FUNDAMENTAL GOALS

Are there any global issues or goals which should shape our thoughts and decisions? Lynch [2] proposed two 'meta-criteria' of urban design. They were justice and efficiency. We have discussed some of the issues associ-

224

ated with equity and access in Chapter 4 but how do we consider efficiency? What criteria can we use to measure it and over what timescale can we consider the impact of our decisions today?

One of the fundamental issues which has emerged over recent years is that of **sustainability** in our decisions and it links directly to the effects and efficiency of our activities. The definition of sustainable human activity of the Bruntland Commission [3] is mentioned in Chapter 4 and it relates to fundamental characteristics of equity, safety, robustness and access.

The quality of our surroundings is not solely related to its equity and sustainability, though these are important prerequisites. The variety and vitality of places, their economic viability, their safety and sensory delight are also important. Places may possess great spatial, visual and experiential qualities, and it is the combination of these which contributes to the character of really great places.

The idea that places play host to a 'community of strangers' has been suggested by several authors notably Jane Jacobs [4] and Barrie Greenbie [5]. This is an important issue to consider because, as we have seen in previous chapters, places play host to a great variety of users. It is important to recognize the great diversity of needs and aspirations which people have for different localities. Not only that; it is also important that we recognize the different sorts of functions and purposes which different places have. Greenbie [5] suggested two clear types of place, each with its particular sorts of needs and characteristics. He used the terms 'proxemic' and 'distemic' to distinguish them. Simply, they relate to the differences between the environment of local communities and those of cosmopolitan spaces.

These definitions, though apparently simplistic, do help us to conceptualize the nature of each sort of place and the differing requirements and opportunities which they present. They also enable us to think about the roles and needs of other sorts of place including those which fall somewhere between the definitions suggested.

Greenbie [5] suggested that the design and management of distemic places should 'find and express symbols that can be shared by people not as separate cultures, but as members of the human race'. He saw these places as having to be intellectually and spiritually accessible to people from diverse backgrounds and origins. He also suggested that proxemic places had very different needs which related much more directly to the local residents, their need for familiar cultural symbols, and the potential for a less obvious structure because they grew to know their own area intimately.

APPRECIATING THE DIVERSITY OF PLACES AND CULTURES

CREATING AND INNOVATING

Many great places and buildings are the result of great creativity, innovation and design. While recognizing the importance of careful planning and participation it is vital that we recognize the importance of vision and ambition. This is well illustrated by the example of two internationally recognized great places, the Pompidou Centre in Paris (Figure 10.1) and St Mark's Square in Venice, which was referred to in Chapter 3 (Example 10.2).

EXAMPLE 10.1

THE POMPIDOU CENTRE, PARIS

Conceived in the early 1970s, this cultural centre has excited great contention, love and hatred. It sits in the traditional street pattern of Paris as a strange and surprising feature with brightly coloured structure and services exposed to full view. Glimpses of the building from surrounding streets contrast with the open plaza, modelled on Sienna's Compo. As a national museum for modern art and centre for industrial design, the building and spaces around it have created a major international place [6].

Figure 10.1 The Pompidou Centre, Paris.

EXAMPLE 10.2

ST MARK'S SQUARE, VENICE

Today one of the greatest attractions for international tourists, this superb square has been the centre of Venetian public life for over a thousand years. Napoleon called it the 'drawing room of Europe'. The square, or piazza, has been redesigned and recreated over the centuries and the essential qualities of space and vitality have endured. Major public buildings and monuments enclose and define the space, which enjoys shaded arcades and open-air cafés. The fine spaces and the relationship to the Canal San Marco provide a setting for public life and activity on a grand scale [7].

CREATIVITY AND VISION

There are many examples of creativity and vision in both buildings and places. Select two examples for study, a building and a place, then identify:

- the main challenges presented in each situation;
- the key ingredients of the response and vision;

- the main actors and how they achieved their objectives;
- any problems, sacrifices or losses which were experienced.

STAKEHOLDERS IN THE BUILT ENVIRONMENT

We all have a stake in the process of urban change. For some, that may seem to be rather a small stake, but we all depend on our town for our way of life; for some, their livelihood revolves around the mechanism of change. For all of us the level of involvement changes over time and we may move from being passive consumers to active shapers in the urban design process. The active participants are well known: the building developers, builders, designers and clients. Some are rather less directly but no less importantly involved, including planners, ground landlords, tenants and so on. We are all affected by change to the urban fabric: when a familiar landmark is demolished, to be replaced by a new construction, we may feel the loss of the familiar and, for a time thereafter, may dislike or at best feel ambivalence towards the new feature. Is this why many people take some time to assimilate and appreciate new urban forms and architectural styles?

People intimately affected by the process of urban change, but hardly ever considered, are the future generations of occupants and users. All present populations have mixed reactions to the heritage of urban forms – from plan form to building style – and we should expect future generations to have the same reaction to our own products. Some forms endure, others are swept away. Some are designed to endure, others are ephemeral. All are products of our own times and society – and, as society changes, its values also change. Can we plan for the future? Perhaps the most important contribution that we could make is the increasing realization that our assets are finite. Should we should design for sustainability, flexibility, robustness and re-use, in order that the future may gain continuing benefit from our creations, albeit in new and adapted forms?

227

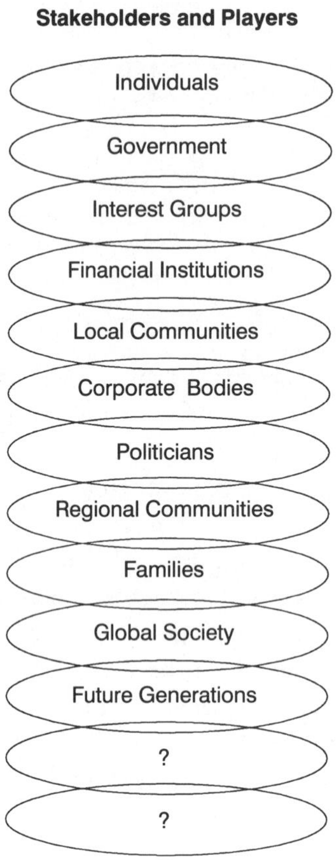

Stakeholders and Players

Figure 10.2 Stakeholders and players.

WORKPIECE l0.2

STAKEHOLDERS IN BUILDING NEIGHBOURHOODS AND PLACES

Who are the stakeholders in the processes of change in the built environment?

The list in Figure 10.2 was compiled in a few minutes. What other interests or actions could be added to the list? Make short notes on the main interests of each type of action or group of actions. Try to identify ways in which different groups have complementary or conflicting interests.

PLACES AND THE DEVELOPMENT PROCESS

The development process can be described very simply as the way in which building and development takes place, the stages of activity involved and the range of 'actors' in the process. This simple definition disguises the complexity of the processes involved and 'the uneven and volatile nature of development'[8]. The ways in which development occurs or fails to occur is important to the character and capabilities of places and it is essential that we try to understand the forces at work.

In Britain the development process has 'been boom-prone and crisis-prone, experiencing much more dramatic fluctuations than the economy as a whole'[9]. This phenomenon is seen as a major problem for the development and construction industries which are themselves liable to under-capacity and high costs, or over-capacity and frequent bankruptcies. Various authors have studied the development process describing the stages in the process and the main actors involved [9]. Example 10.3 describes the following key stages and 'main' actors.

EXAMPLE 10.3

STAGES AND ACTORS IN THE DEVELOPMENT PROCESS

Stages	Actors
* Initiation	* Landowners
* Evaluation	* Developers
* Acquisition	* Statutory bodies
* Design and costing	* Financial institutions
* Permissions	* Building contractors
* Commitment	* Professional advisers
* Implementation	* Objectors
* Letting, managing or disposal	

Derived from Cadman, D. and Austin-Crowe, L. (1991) *Property Development* [9].

Other authors have pointed out that these processes do not occur in isolation but that they are subject to broader social, economic and political forces; and that the forms which development takes can therefore vary greatly in different places and at different times. It has been suggested that there are two components of the complex interactions which may occur. The first is **agency,** or the way 'in which actors in the development process define and pursue their strategies, interests and actions'. The second is **structure** or 'the socio-economic and cultural framework' within which actors define and pursue their strategies, interests and actions [10]. These definitions may seem to be similar but the important point is the complexity of motives generated by individual and corporate agents and the influences upon them of varied and changing social and cultural contexts which may occur.

Recognizing the complexity of the forces at work and the values and attitudes which underlie the development process is important to our understanding of changes in the built environment. Can it help us to make connections between actors and events in the process and to recognize the influence of these on the form and use of the resultant places and buildings?

WORKPIECE 10.3

STUDYING THE DEVELOPMENT PROCESS

Find a small building project that has just been finished.

- Try to find out how long it took from inception to completion, and make a calendar of events showing the main stages.
- Make a list of all the types of people who were involved and show when they were involved on the calendar.
- Were there any conflicts of interest?
- How is the finished product regarded by the people involved?

DEVELOPING SHARED AGENDAS

Fundamental to all changes in urban form are the values which underpin them. Clearly, we do not all hold exactly the same values: indeed, it would be very boring if we did! Nevertheless, there may be a range of characteristics which we all generally welcome and enjoy. These may include the warmth of a summer's day, the beauty of a clear starry night, or the tranquillity of a 'sacred' place. Could these qualities be better preserved or achieved if we all sought to understand and share an understanding of what they are? This is quite an idealistic notion. It is not intended to be so, but the importance of our moral values in both shaping and using urban places should not be under-estimated.

DEVELOPING SHARED OBJECTIVES

In the crowded urban environment, the development of any site will affect the surroundings of many people, meeting some needs but causing problems for others. If urban design is to spread benefits more equitably, it must solve this problem by addressing two points. It should be inclusive, in that it should seek to include as many of the potential users of the development as possible in the design process. It should also be enabling, in that these users are empowered to meet their own needs without compromising those of others, or unnecessarily having to resort to costly professional assistance for every intervention. Various design and development processes have different degrees of including and enabling, and are more or less successful in developing shared objectives.

In the UK the planning system has limited scope for identifying or involving all those affected by development. It tends to consult (or inform) rather than to include in the decision-making process. This approach is often channelled through agencies which see themselves as

230

providers rather than enablers of development. Other examples have been developed in various countries, though they are still the exception rather than the rule.

Social landscapes are representations of the issues raised as important by a community, and its views of possible remedies. Based on a number of complementary techniques, including meetings, interviews and questionnaires, this method seeks to be inclusive by encouraging everyone to participate. The information gained forms the basis of proposals from which decision makers can act. In this way, the community is the source of ideas and the professional is a facilitator. This approach was tested extensively by John Donovan as research into equitable urban design, and is explained in Chapter 4.

Everyone who studies the built and natural environment, or who seeks to influence its future conservation or development, should consider the wide-ranging communities of interest which exist and which should be understood and respected in the decision-making process and action. As we have suggested earlier, some communities are better equipped or able to participate in these quite complex and sometimes obscure decision-making processes, and indeed some interests may seek to avoid or frustrate participation. Being aware of the conflicting interests, and the relative influence which different interests may be able to muster, is an asset for all professionals involved with shaping the future of our buildings, neighbourhoods and places. Does everyone recognize the privilege which this understanding gives them? Do some abuse their position to promote personal or sectoral advantage?

For local communities to be able to represent their interests effectively, they need to possess knowledge, time and skills. Is it possible that some groups could abuse these in competition with other legitimate interests? Do they need to develop and possess values and attitudes which support mutual benefit and achievement rather than parochial advantage and rivalry? Blackman has said that 'in essence, community development is about people taking action together and developing the knowledge, skills and motivation to express their needs and improve conditions, either in a geographical neighbourhood or for a particular 'community of interest' [11].

The idea of geographical communities with 'local' interests, and communities of interest comprising people with shared or common aspirations, is important. The former may directly affect the approach to particular places, while the latter may influence the way we approach issues globally.

PROFESSIONALS AND COMMUNITIES

UNDERSTANDING PLACES

In earlier chapters we have shown some of the influences that the natural environment and people's past aspirations have had upon the form, use and quality of the built environment. By thinking about the forces and events which have led to the present character of places, we can put ourselves in the best position to see how they could be improved in the future.

One of the characteristics of less mobile times was that people had a long-standing knowledge and understanding of their surroundings. In more recent times people have moved from their place of birth and may have travelled widely since. Does this mean that their understanding of one place and what makes it 'tick' is lost? Professional practice and investment can now spread over the whole of the globe, bringing the risk that the distinctive local conditions and circumstances will be ignored in favour of simplistic, transferable international 'solutions'.

Understanding the forms of places as they are and how they came to be as they are is essential before we try to develop visions or designs for the future. This understanding and a high level of sensitivity are needed at every level from the development of a small site, through to urban design strategies for whole cities and regional planning.

The Birmingham Urban Design Study [12] described in Chapter 6 is of interest because of its wide spatial remit, being one of a growing number of area-based studies and incorporating many 'quarters' of clearly differing character. In its presentation of concept and precedent, key design and development issues are set out, without resorting to a design guide *per se*, or setting down rules which might prejudice future detailed decision making. Although many of its suggestions are rooted in the visual aspects of urban design – new façades, landmark buildings on key sites and so on – it clearly also pays regard to many other urban design criteria including pedestrian accessibility and safety, landscaping and urban vitality, all without constraining the future shapers of the urban form of Birmingham. A pedestrian movement study provided an invaluable complementary strategy for diminishing the dominance of roads and cars upon the quality of the city core [13].

SHARING EXPERIENCE AND EXPERTISE

Expertise takes time to develop and may not be equally available. Spreading knowledge, skills and expertise more widely and sharing experiences is valuable for us as individuals and communities. Many efforts are being made to provide support and communication systems.

A few examples include:

● know-how networks;

- environmental forums;
- Euro-cities networks;
- common purpose;
- centres for design and architecture.

EXAMPLE 10.4

DESIGN AND ARCHITECTURE CENTRES

The concept of architecture centres is demonstrated in a variety of examples from Europe and North America: for example, the Swedish Museum of Architecture in Stockholm, the Pavilion de l'Arsenal in Paris and the Design Exchange in Toronto. It is not that the enterprise needs a centre – Philadelphia has promoted a range of activities and networks without the need for a permanent centre. The essential idea is that good design, planning, environmental care and community involvement are promoted and enabled through support and action

programmes and, usually, a centre of activity. The actual emphasis of different centres vary considerably. Studies in the UK [13] have identified seven potential activities:

- creating networks;
- informing for participation;
- museums;
- pressure group;
- promoting debate;
- education;
- presenting architecture and design.

Figure 10.3 Pavilion de l'Arsenal, Paris. (David Chapman Jr.)

WORKPIECE 10.4

NETWORKS AND FORUMS

Do any networks, forums or help groups exist in your area? Try to identify some that do and list their strengths and weaknesses.

IMPLEMENTING SHARED AGENDAS

Although the processes of developing and sharing ideas and visions for the future of places is very important, it is not the end of the process. It is the beginning. So many good intentions and exciting ideas fail to become reality. Why is this? How can we improve the chances or probability of success at the small and the large scale?

ENABLING AND FACILITATING

One of the significant changes in the way public policy-makers, and particularly planners, have begun to operate in recent years is the shift from regulatory modes to emphasis upon 'enabling and facilitating'. This change is one of ethos and intention. There may always be some need for strategic and local planning together with the control and guidance of development within the context of some broad framework. What has changed is the growing recognition that positive action and enabling is needed in many instances to facilitate the realization of public aspirations and planning policies. The important question raised by Stoker and Young – 'Who benefits?' – is fundamental to every context [14]. As they asked, 'Do you plan on the basis of what the private sector can be induced to deliver, letting it largely determine the patterns of land use? Or do you analyse local social conditions and plan on the basis of tackling local needs?'

Many inner-city and industrial regeneration programmes have been supported by entrepreneurial planning. Adams describes this as 'quite different from the traditional planning activities of development plan-making and development control. It is action orientated or implementation centred rather than process orientated'[8]. Characteristics of entrepreneurial planning include preparing urban design 'visions', supportive grant and finance packages, land assembly, partnership schemes, education and training 'compacts' and any other intervention which could legally overcome obstacles or facilitate action.

Like all entrepreneurial activities there are potential risks. Financial packages can fail. Well-intentioned community involvement can backfire. 'Visions' can founder as only fragments of them are achieved.

As we have suggested in earlier chapters, the natural environment and the historical infrastructure of places exert a powerful influence on their future development. In the same way it may be possible for new infrastructures to have very significant influence upon development and investment decisions. These effects may be deliberately contrived to stimulate regeneration or may be more accidental spin-offs of infrastructure installed for quite different reasons.

INFRASTRUCTURE AND DEVELOPMENT

The concept of city centre management, introduced in Chapter 6, is emerging in the United States and Britain as a way of revitalizing older retail centres, partly because of the threat from out-of-town retail centres and partly as a recognition of the value of bringing together all of the public and commercial interests, not only to encourage good physical places but also to promote lively activities, celebrations and a carefully constructed approach towards servicing, access, maintenance and management. Making these connections between different interests and agencies facilitates shared objectives and collaboration action. Could the principles be transferred to various other settings?

CITY CENTRE MANAGEMENT

Many people feel powerless in influencing the changes which seem to be going on around them. Others by personal leadership qualities, economic influence or democratic power can have very significant influence. The privilege of influence requires that we consider carefully how the power it confers is exercised. What values and attitudes underpin our thoughts and actions? People, politicians and professionals can all consider where they stand and how their leadership and participation will affect the qualities of the living environment. Will they foster partnership and collaboration? Will they promote harmony or discord?

THE NATURE OF INFLUENCE

Shaping the future of the built environment may be seen to be concerned with participation and management as much as design. These activities occur on different spatial and temporal scales. Appreciating the many contributions made by the different actors in the processes of management and change is an important precursor to individual action and collaboration.

INTERDISCIPLINARY COLLABORATION IN URBAN DESIGN AND MANAGEMENT

It has been suggested that 'the future of urban design, indeed the future of a rewarding urban experience, is in the hands of ... traffic engineers, landscape planners and cultural managers' [15]. This is a significant observation, and one which must be understood if we are to have any influence on the future, not by individually commanding centre stage but by recognizing the important contributions which many actors and agents of change exert on our environmental and social welfare. It requires a degree of humility and may influence policies for practice and education. Will it eventually influence the alignment of professional bodies as they currently exist?

What are the values and attitudes which affect our actions? What are the generic skills which are needed and what specialist skills are also needed? Perhaps the most important question is how can the processes of urban policy, design and management be made accessible to people and equitable to all? Adversarial stances and arrogant claims are anathema to participation and collaboration. Instead there are a lot of questions which we should ask ourselves, about our values, where we stand and what part we want to play.

As Goodey warned;

> In approaching the year 2000 we must ask whether we can exert any significant impact on the fortunes of our cities and whether this impact recognizes economic and human motives ... or tries, instead, an imposed doctrinaire solution.[15]

In describing the award-winning scheme for Victoria Square which followed from the Birmingham Urban Design Study mentioned earlier, a city council officer explained:

> I cannot stress too highly the importance of good interdisciplinary working relationships and understanding of respective responsibilities at the various stages to achieve a successful outcome on complex projects.[16]

Considering and managing change in the built environment is not an exclusive expertise possessed by a chosen few, but the values and attitudes which underlie the processes of change have broad relevance and applicability to everyone concerned with the equity and quality of urban environments. They require public involvement, interdisciplinary collaboration and shared agendas for future change and quality.

CHANGE AND THE FUTURE

The speed of change and development seems to be accelerating almost inexorably. Whether this acceleration is real or imagined, it is certain that the future will in some ways be unimaginably different from today, but perhaps in many ways it may be quite the same. This apparent contradiction may be explained by the combination of basic human needs and motivations with changing capability and potential.

Every technological innovation potentially transforms our human capacity. The development of early bronze and steel tools accelerated the clearance of woodland for cultivation. Today the explosion in communication and information technology looks likely to transform our life styles very rapidly. How this transformation will occur, what it will mean for individuals and communities and how fast change will take place are all imponderables. We can speculate but we cannot be sure.

One characteristic of change is the probability that there will be a time lag between the first sight of possible change and that change actually becoming endemic or systemic. This is well demonstrated by science fiction novels which may speculate about future possibilities years before they become practical possibilities. Arthur C. Clarke, H. G. Wells and Jules Verne are classic examples, while William Gibson and others have more recently speculated on the consequences of information technology and, *inter alia*, its impact on urban form and behaviour.

Each technological development is in effect a 'tool' which extends our limited human capabilities. Whether the ultimate effects of this new power will be benign or malignant may be unknown. The invention of automobiles and tractors have both had significant impacts upon the lives of many people. Both have opened up new possibilities but will each have different ultimate consequences?

In the late twentieth century the development of communication and information systems based upon silicon chips, fibre optic cables and satellites seems set to change our lives in profound ways. Will the effects of these new capabilities really change our lives or merely be an incidental component?

The development of virtual reality systems seems to offer us the possibility of every experience we desire in the comfort of our own homes! If so, we could solve the problem of traffic congestion very quickly. But would such a virtual reality really change our behaviour and can it really satisfy our needs? The potential to simulate proposed developments, existing places or past periods through virtual reality is certainly likely to transform our ability to visualize the future [17]. Will it change our attitude to the conservation of the past?

CONCLUSION

This book has taken a broad and synoptic view of the forces at work in the built environment; the forms of buildings and settlements which result; and the ways people occupy and use them. Some people may see the wide range of issues raised as only remotely connected to their individual concerns and responsibilities but we should all see that individual attitudes and actions have an effect upon the shape and quality of our living environment.

Part One explored the forces which led to development in some places rather than others; the influence of the natural environment upon the form of settlements and buildings; the impact of human aspirations and capabilities upon the growth and decline of places; and the resultant form and use of those places.

Part Two considered the ways in which changes in the built environment and the resultant settings are experienced and appreciated by their occupants. The basic qualities of safety and equity for current and future occupants of places is an essential prerequisite for enjoyment of the aesthetic and sensual qualities of place and space. The vitality and variety of uses, experiences and opportunities are inextricably part of the totality of experience of place.

Part Three explored some of the ways that public policies and actions influence the qualities and development of places, how users and stakeholders can be involved or excluded from the decision-making process, and how local communities of interest can represent their aspirations and achieve their goals. The importance of appreciating the diversity of interests, careful analysis of the qualities of places, and involvement in developing shared visions and agendas is highlighted together with the importance of making connections across areas of knowledge and between a diversity of interests.

WORKPIECE 10.5

PUTTING IT ALL TOGETHER

All of the themes explored in this book are quite simple. They embrace a whole range of disciplines and specialisms which individually we may pursue, but how could our specialist skills combine to create vital and varied buildings, neighbourhoods and places?

Draw up a checklist or a matrix of the key factors which you think could underpin the activities of all of our individual or professional actions. These ideas could be used for discussion with colleagues and they can be reviewed regularly.

USING AND CELEBRATING

The physical form of our towns and villages gives us the stage-set. We are the players. Our activities as residents or visitors transform the inert place into a vital experience. The ways in which we use and manage public places are, therefore, of great importance. But do not forget the creative and innovative qualities of great architecture and urban design!

All cultures have traditions and ways of celebrating events, seasons and achievements. Some are more exuberant, or even more odd, than others, but they all add vital highlights to the daily routine of activities. The summer carnival, the torchlight procession, the street market and fair bring people together in public places and, in Cullen's words, 'drama is released'[18].

Figure 10.4 Stockholm street life.

Figure 10.5 Café, Mdina, Malta.

Figure 10.6 The Grand Canal, Venice.

However, we would not last long if these celebrations went on continuously. The simple uses of our public places – walking or promenading, meeting and resting, or window-shopping – all add to the life of the place. Providing appropriate and enjoyable settings for these activities is important, and Lynch urged us to observe existing behaviour settings both as a means of interpreting the character of existing places and to help us form new and enjoyable settings [2].

We have explored a range of historical, theoretical, physical and emotional facets of the urban design process. We have also touched on some of the tools and techniques which might help to appraise the process and, possibly, influence it. As the authors of Responsive Environments [19] and Francis Tibbalds [20] suggest, it is the way in which we 'put it all together' that is of paramount importance. Individually, none of us can control or direct the pattern of change, no matter how much power or influence we may fleetingly obtain. Yet we can all make a positive contribution to the process, no matter how modest that contribution may seem to be. Each individual, using the public space of our settlements positively and thus contributing to its vitality, character and safety, is adding to the quality of life of that place. Each shopkeeper contributing to the street scene, and each bank supporting public art, adds to our enjoyment. Every engineer and transport company adding to the accessibility of places and the user-friendliness and usability of places does the same.

Effective means of guaranteeing safety for residents and visitors alike underpin all of our efforts to create physically attractive and enjoyable public spaces. Every effort which promotes public life and space which is equally accessible to all adds to the quality of life of us all. Every action which undermines these efforts, or privatizes the public domain, deprives us all. At worst they could foster civil disorder and protest as seen in several cities in recent years.

It would be easy to become pessimistic about the failures of recent years in terms of planning and civic design. Somehow, after two world wars, we seemed to have lost our way. In the determination to create a new world, we lost sight of the old one. But we are also learning to seek actively the processes of design that increase equity and guide the physical form and activities of our settlements to have the qualities which we most need and cherish. The concept of urban design is far larger, more wide-ranging, than many have implied in their examinations of processes and techniques of changing the physical fabric. As Bacon asserts, 'since designers should provide a setting for a totally harmonious life experience, the dimensions of their designs should encompass the whole of the day, the whole of the city' [21].

We have raised a range of issues and suggested a number of tools which we may use. The challenge is there for us: to question our individual actions; to contribute towards shared objectives for our local and global community; to adapt and develop the tools at our disposal to respond to the changing problems we face; to create lively and beautiful neighbourhoods and places!

SUMMARY

The design of buildings, planning for new developments and the management of urban places involves complicated and interrelated activities. The smallest individual decision or action could have an impact upon the character and qualities of places. Understanding the forces at work and the forms and uses which places support is important for both analysis and and intervention.

Throughout the book we have:

- seen why an understanding of the forces which influence the shape of the built environment is important;
- developed the ability to reflect upon the types of influences at work and the possible ranges of outcomes from those forces in both physical form and use of buildings and places;
- considered the qualities and opportunities which built environments can provide for their users;
- raised questions about the values and attitudes which underpin the actions of actors and agents in the planning and development process;
- recognized the importance of creativity and innovation in cultural development and problem solving activity.

CHECKLIST

The issues addressed in this book are complex and fundamental to everyone involved with shaping the built environment. The checklists are not intended as a prescription for every action but as an aid to consideration. Each situation we confront deserves individual attention. What will be the consequences of different courses of action? Who will win and who could lose? Which course will contribute to the probability of beneficial outcomes for people in the built environment, now and into the future?

- The places we experience are the result of the complex interactions between the natural environment and human interventions.

- The character of places derives from the physical form of places and the way people are able to use them.
- Each place provides a range of settings which may or may not enable its users to satisfy their needs.
- Patterns of development, infrastructure and distribution of facilities have direct and important implications for equity and access for different occupants of a place.
- Involving people in the decision-making processes of change can help to create more responsive and democratic places.
- It is possible to use various techniques to study the 'evolution' of places and to explore different strategies for the future.
- There are several examples of successful participation in the process of developing 'shared visions' for the future of local places.
- Everyone, including professionals and students of the built environment, can contribute to building successful neighbourhoods and places. This requires sensitivity, willingness to collaborate, and values and attitudes which promote involvement and participation.
- Buildings are important for both comfort and cultural reasons; they consume resources and energy and care should be taken to maximize their benefits and minimize their energy consumption.
- The built environment is a manifestation of the culture of the society which creates it. Creativity, innovation and cultural development are vital components of a lively and healthy society.

REFERENCES

1. Stein, C.S. (1966) *Towards New Towns for America*, MIT Press, Cambridge, Massachusetts.
2. Lynch, K. (1981) *Good City Form*, MIT Press, Cambridge, Massachusetts.
3. World Commission on Environment and Development (1987) *Our Common Future*, Oxford University Press, Oxford.
4. Jacobs, Jane (1961) *The Death and Life of Great American Cities*, Random House, New York.
5. Greenbie, B. (1984) Urban design and the community of strangers. *Landscape Design* **149**, 8–11.
6. Knobel, L. (1985) *The Faber Guide to Twentieth Century Architecture: Britain and Northern Europe*, Faber and Faber, London.
7. Everyman Guide (1993) *Venice*, David Campbell Publishers, London.
8. Adams, C.D. (1994) *Urban Planning and the Development Process*, UCL Press, London.
9. Cadman, D. and Austin-Crowe, L. (1991) *Property Development*, 3rd edn, E & FN Spon, London.

10.   Healy, P. and Barrett, S. (1990) Structure and agency in land and property development processes: some ideas for research. *Urban Studies*, **27**, 89–104.

11.   Blackman, T. (1995) *Urban Policy in Practice*, Routledge, London.

12.   Tibbalds Colbourne (1990) *Birmingham Urban Design Study*, City Council, Birmingham.

13.   Coonan, R., Cowans, R. and Davy, P. (1993) Architecture Centres. *Architectural Review*, **CXCII** (1154), 65–73.

14.   Stoker, G. and Young, S. (1993) *Cities in the 1990s*, Longman, London.

15.   Goodey, B. (1993) Urban design of central places and beyond, in Hayward, R. and McGlynn, S. (eds) *Making Better Places: Urban Design Now*, Butterworth, Oxford.

16.   Wright, G. and Blakemore, J. (1995) Victoria Square, Birmingham. *Urban Design*, **54**, April, 21–24.

17.   Hall, T. (1988) Computer visualisation for design control. *The Planner*, **74**(20), 21–25.

18.   Cullen, G. (1961) *Townscape*, Architectural Press, London.

19.   Bentley, I. *et al.* (1991 edn) *Responsive Environments: a manual for designers*, Architectural Press, London.

20.   Tibbalds, F. (1992) *Making People-friendly Towns*, Longman, Harlow.

21.   Bacon, E. (1992 edn.) *Designing Cities*, Thames and Hudson, London.

FURTHER READING

Blowers, A. *et al.* (1982) *Urban Change and Conflict*, Paul Chapman, London.

Chapman, D. and Larkham, P. (1994) Under*standing Urban Design: an introduction to the processes of urban change*, UCE, Birmingham.

Crosby, T. (1973) *How to Play The Environment Game*, Penguin, Harmondsworth.

Day, Christopher (1990) *Places of the Soul*, Harper Collins, London.

Devas, N. and Rakodi, C. (1993) *Managing Fast Growing Cities*, Longman, Harlow.

Elkin, T. and McLaren, D. (1991) *Reviving the City*, Friends of the Earth, London.

Owen, S. (1991) *Planning Settlements Naturally*, Packard Publishing, Chichester.

Talbot, M.(1986) *Reviving Buildings and Communities*, David and Charles, Newton Abbot.

INDEX